Colloidal Nanoparticles

Functionalization for Biomedical Applications

Colloidal Nanoparticles

Functionalization for Biomedical Applications

Nikhil Ranjan Jana

CRC Press
Taylor & Francis Group
Boca Raton London New York

CRC Press is an imprint of the
Taylor & Francis Group, an **informa** business

CRC Press
Taylor & Francis Group
6000 Broken Sound Parkway NW, Suite 300
Boca Raton, FL 33487-2742

Printed on acid-free paper

International Standard Book Number-13: 978-1-138-33760-2 (Hardback)

Library of Congress Cataloging-in-Publication Data

Names: Jana, Nikhil Ranjan, author.
Title: Colloidal Nanoparticles : Functionalization for Biomedical Applications / Nikhil R. Jana.
Description: Boca Raton : CRC Press, Taylor & Francis Group, 2018. | Includes bibliographical references and index.
Identifiers: LCCN 2018053997| ISBN 9781138337602 (hardback : alk. paper) | ISBN 9780429165603 (ebook)
Subjects: LCSH: Biomedical materials. | Nanostructured materials.
Classification: LCC R857.M3 J36 2018 | DDC 610.28--dc23
LC record available at https://lccn.loc.gov/2018053997

Visit the Taylor & Francis Web site at
http://www.taylorandfrancis.com

and the CRC Press Web site at
http://www.crcpress.com

Contents

Preface

The primary motivation for writing this book comes from my personal experience in training graduate students and interacting with nonchemist professionals during my collaborative research. I found many common mistakes and misconceptions from their side in selecting the right experimental procedure from the vast body of scientific literature.

This book consists of seven chapters, viz., the application potential of nanoparticles and importance of functionalization, design considerations of nanoparticle functionalization, chemical synthetic methods of selected nanoparticles, selected coating chemistry for transforming a synthesized nanoparticle into a water-soluble nanoparticle with a cross-linked shell, selected conjugation chemistries for making chemically/biochemically functionalized colloidal nanoparticles, the state of the art and future of clinic nanodrugs, and finally guidelines to solve common issues faced in preparation of functional nanoparticles.

Most of the contents of this book are from different published articles from our group as well as from other selected research groups. I am very thankful to those publishers and authors from whose publications I have collected the useful information.

I hope this book may provide guidance to graduates and researchers who are beginning their careers in experimental nanoscience. In particular, this book will provide experimental protocols to chemists/materials scientists/biologists and also guide graduates of other disciplines. It could also be relevant to nanobiotechnology courses.

Professor Nikhil Ranjan Jana
Indian Association for the Cultivation of Science
Kolkata 700032, India

Acknowledgments

I would like to thank Professor Tarasankar Pal, Professor Catherine J. Murphy, Professor Xiaogang Peng and Professor Jackie Y. Ying, with whom I have had a long association during my research career. I would also like to thank my research group members Dr. Nibedita Pradhan and Suman Mandal for their support in writing this book.

Author

Nikhil Ranjan Jana is a professor at the School of Materials Science, Indian Association for the Cultivation of Science. He received his undergraduate degree (1987) from Midnapore College and his masters (1989) and PhD degrees (1994) from the Indian Institute of Technology, Kharagpur. Prof. Jana worked as a postdoctoral fellow at the University of South Carolina (1999–2001) and University of Arkansas (2003) and as a scientist at the Institute of Bioengineering and Nanotechnology in Singapore (2004–2008). His research group focuses on chemical synthesis of functional nanoparticles/nanobioconjugates, the design of nanoparticles to control cellular processes (e.g., endocytosis, autophagy, apoptosis), and the development of nanoprobes/nanodrugs for subcellular targeting/imaging and inhibiting amyloid aggregation under intra-/extracellular space. His group has published 150 peer-reviewed research articles in internationally recognized journals, which have about 20,000 citations.

CHAPTER 1

Application Potential of Nanoparticles and Importance of Functionalization

1.1 INTRODUCTION

The biomedical application of nanoparticles is a rapidly emerging research area. This is because nanoparticles have length scales similar to various biomolecules and biological events and they can be designed to understand the language of the biological world. Although the size-dependent property of many nanoparticles is well established, their transformation into nanoprobes that can provide information about the biological world is challenging. This chapter focuses on the importance of the nanometer length scale and the role of functionalization of nanoparticles in the context of their biomedical application. We have selected only a few nanoparticles to demonstrate this biomedical application potential. The selected nanoparticles have fluorescent/plasmonic/

magnetic properties and are widely used as optical biosensors, imaging probes, phototherapy agent and drug delivery carriers.

1.2 WHY NANOPARTICLES?

The nanometer length scale is just above the molecular length scale and typically spans between 1 and 100 nm. This length scale is unique, as material properties are size dependent at this length scale.[1,2] For example, metal to nonmetal transition occurs at the 1–5-nm length scale, the emission color of CdSe semiconductors depends on the size of particles in the 2–6-nm range and the plasmonic property of silver (Ag)/gold (Au) depends on its size in the 5–100-nm range.[1,2] A variety of materials are synthesized at this length scale and are commonly called nanoparticles. They include semiconductor nanoparticles that are conventionally known as quantum dots (QDs), metal nanoparticles, metal oxide nanoparticles, metal sulfide nanoparticles, polymer nanoparticles, polymer micelles, liposomes, vesicles and so on.[3–8]

From the biomedical point of view, the nanometer length scale is unique, as materials science meets biology at this length scale and nanoparticles can be used to understand the language of the biological world.[9–13] In particular, there are many biological entities at the nanometer length scale, and many biological events occur at this length scale. Examples include the size of globular proteins, self-assembled biomembrane structures, molecular recognition events involving multiple chemical functional groups, DNA replication, inorganic-organic nanocomposites of bone material, cellular endocytosis and so on.[9–13] As synthetic nanoparticles have length scales similar to many biomolecules and biochemical events, they are very useful to study/monitor/control many biological events.

1.3 WHY FUNCTIONALIZATION?

Nanoparticles interact with a biological interface through their surfaces and in particular through the chemical functional groups present at their surfaces. Bare nanoparticles generally do not exist. Most nanoparticles exist in their colloidal form, and in some limited cases, they exist in powder form or composite with other solid substrates. In colloidal form, the surface of nanoparticles is covered with surface-adsorbed ions/molecules/polymers that provide their colloidal stability. Thus, adsorbed ions/molecules/polymers critically dictate the interaction with biological interfaces, and if we want to control this interaction, we actually need to control the nanoparticle surface chemistry. The functionalization of nanoparticles refers to the chemistry involved in preparing nanoparticles with designed surfaces decorated with specific chemicals/biochemicals.[14-21] First, we need to know the commonly occurring surface structures of nanoparticles. Figure 1.1 demonstrates the typical surface chemistry that commonly exists. In one such class, nanoparticles are capped with small molecules, ions or polymers. Examples of such particles include hydroxide-capped metal oxide nanoparticles, citrate-capped Au nanoparticles and hydrophilic surfactant-capped metal/metal oxide nanoparticles. In the second class, nanoparticles are capped with hydrophobic surfactant where the surfactant forms a compact self-assembled monolayer structure.[22-26] This class of

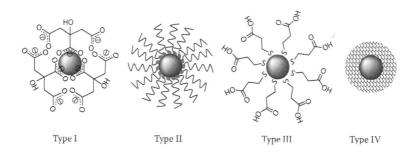

| Type I | Type II | Type III | Type IV |

FIGURE 1.1 Common types of surface chemistries of nanoparticles. In type I, nanoparticles are capped with small molecules, ions or polymers; in type II, nanoparticles are capped with surfactant with self-assembled monolayer structure; in type III, nanoparticles are capped with molecules via strong chemical interactions; and in type IV, nanoparticles are coated with cross-linked polymer shells.

nanoparticles has been the most widely studied in the last 25–30 years because of their high-quality, large-scale synthesis options and because they can be processed like a chemical compound. Examples include hexadecylamine-capped CdSe nanoparticles, dodecanethiol-capped Au/Ag nanoparticles and oleic acid-capped iron oxide nanoparticles. These nanoparticles are soluble in organic solvent, but not in water, because of the hydrophobic nature of the capping surfactant. The third class of nanoparticles are capped with molecules/polymers with relatively strong interaction via specific chemical interactions.[27–30] These classes of nanoparticles have been developed since the emergence of the second class of nanoparticles for their more effective utilization. Examples include mercaptoundecanoic acid-capped Au nanoparticles, lipoic acid-capped QDs, dopamine-capped iron oxide nanoparticles and polymeric thiol-capped QDs.[15,27–30] Limited surface chemistry and chemical reactions can be exercised using this type of capped nanoparticles. The fourth class of nanoparticles are capped with cross-linked polymeric shells.[21] In this case, nanoparticles are well protected from the external environment, and extensive surface chemistry can be used, which is required for their functionalization. Examples include silica-coated nanoparticles, polyacrylate-coated nanoparticles, nanoparticles with cross-linked dendron shells and shell cross-linked organic polymer-coated nanoparticles.[21]

The surfaces of these nanoparticles need to be modified in order to control their interaction at biological interfaces (Figure 1.2). There are three specific reasons for such modifications.[21] First, water-soluble nanoparticles are required to study biomedical applications. This is particularly because cellular, clinical and other biomedical studies are performed in the aqueous phase. However, most good-quality nanoparticles are capped with hydrophobic surfactant, are water insoluble and require appropriate surface modification to make them water soluble. Second, nanoparticles often induce nonspecific interactions with biological interfaces due to their high surface area associated with multiple surface chemical groups. This nonspecific interaction needs to be

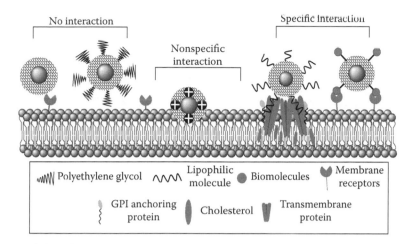

FIGURE 1.2 Importance of functionalization of nanoparticles. As synthesized, nanoparticles either do not interact or interact nonspecifically with the biological interface (e.g., cell membrane shown in the middle). In contrast, functional nanoparticles offer specific interaction with the biological interface.

minimized in order to have specific nano-bio interaction. Thus, appropriate surface modification is required to minimize this nonspecific interaction. Third, specific biological interactions of nanoparticles require appropriate surface modification with specific biochemicals such as vitamins/carbohydrates/peptides/antibodies. Nanoparticles with such surface modifications can be utilized for selective interaction with specific biomolecules or detection of specific biochemical activity. Examples include selective interaction of glycoprotein concanavalin A with glucose-functionalized nanoparticles, selective interaction of galactose-functionalized nanoparticles with galactose receptor overexpressed cancer cells and selective interaction of biotin-conjugated nanoparticles with streptavidin.

There are two specific steps associated with the functionalization of nanoparticles (Figure 1.3). The first step is the coating chemistry. In this step, the nanoparticle surface is coated with

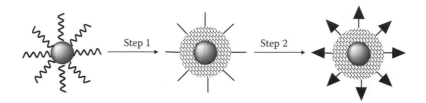

FIGURE 1.3   Steps for functionalization of nanoparticles. In the first step, the nanoparticle is coated with a molecule/polymer that protects it from aggregation and offers chemical functional groups on the surface. In the second step, the coated nanoparticle is covalently linked with the chemical/biochemical of interest using the chemical functional groups at the nanoparticle surface.

an additional molecule/polymer that covers the whole surface. In this process, original molecules at the nanoparticle may be completely or partially replaced with new molecules/ligands. The coating molecule/polymer is chosen in such a way that it provides good water solubility of the nanoparticle and offers chemical functional groups for performing chemical reaction/conjugation. Details of these processes are discussed in Chapter 3. The second step is conjugation chemistry. In this process, nanoparticles are covalently linked with selected biomolecules of interest. Known organic chemistry reactions are used between the functional groups of coated molecules/polymers and the biochemical of interest. Details of these processes are discussed in Chapter 4.

1.4 WHY SELECTED NANOPARTICLES AND SELECTED METHODS?

We have discussed only a few nanoparticles among many reported nanoparticles, and some selected coating and conjugation methods among many reported methods. Such selections are based on the following three reasons. First, although many nanoparticles are reported, only a few of them are most commonly used in biomedical science because of their unique physical/chemical properties. Mostly those nanoparticles are selected here (Table 1.1). Nanoparticles

TABLE 1.1 Summary of Different Nanoparticles That Are Discussed in This Book along with Their Biomedical Application Potential

Name	Composition, Size	Property	Biomedical Application
Plasmonic Au, Ag	Au/Ag, 1–100 nm	Colored	Biosensing/ phototherapy
Quantum dot	CdSe-ZnS, 1–6 nm	Fluorescent	Bioimaging
Metal oxide	Iron oxide/ZnO, 5–50 nm	Magnetic/ fluorescent	Bioimaging, magnetic separation
Doped nanoparticle	Mn-ZnS/MnZnSSe, 4–6 nm	Fluorescent	Bioimaging
Doped nanoparticle	N,F-TiO$_2$, 50–200 nm	Photocatalyst	Phototherapy
Carbon nanoparticle	Carbon, 1–10 nm	Fluorescent	Bioimaging
Graphene, graphene oxide	Carbon, 50–1000 nm	Biocompatible	Biosensing, drug delivery
Hydroxyapatite	Calcium phosphate nanorods, 2–5 × 10–1000 nm	Biocompatible, biodegradable	Drug delivery
Protein	Liposome/albumin, 10–100 nm	Biocompatible	Drug delivery

include two types of plasmonic nanoparticles of different sizes (Au and Ag), one magnetic nanoparticle with different sizes (iron oxide), one QD of different emission colors (ZnS-capped CdSe), fluorescent carbon nanoparticles, two doped semiconductor nanoparticles (Mn-doped ZnS and N, F-codoped TiO$_2$), two oxide nanoparticles (ZnO, TiO$_2$), hydroxyapatite nanorods, graphene oxide/chemically reduced graphene oxide, liposome, albumin nanoparticles and two polymer nanoparticles. Second, although many coating and conjugation methods are reported, only a fraction of them are more powerful and routinely used in generating functional nanoparticles. Those approaches are the main ones selected. Third, while many nanoparticle synthesis methods and coating chemistries are reported, some of them are selective to specific nanoparticles and unlikely to be adapted to other nanoparticles. Most reproducible and generally applicable methods are summarized based on our own experience in developing new methods or adapting the reported methods.

REFERENCES

1. Alivisatos, A. P. 1996. Perspective on the physical chemistry of semiconductor nanocrystals. *The Journal of Physical Chemistry*, 100, 13226–13239.
2. Jain, P. K., Huang, X. H., El-Sayed, I. H. and El-Sayed, M. A. 2008. Noble metals on the nanoscale: Optical and photothermal properties and some applications in imaging, sensing, biology, and medicine. *Accounts of Chemical Research*, 41, 1578–1586.
3. Michalet, X., Pinaud, F. F., Bentolila, L. A., Tsay, J. M., Doose, S., Li, J. J., Sundaresan, G., Wu, A. M., Gambhir, S. S. and Weiss, S. 2005. Quantum dots for live cells, *in vivo* imaging, and diagnostics. *Science*, 307, 538–544.
4. Medintz, I. L., Uyeda, H. T., Goldman, E. R. and Mattoussi, H. 2005. Quantum dot bioconjugates for imaging, labelling and sensing. *Nature Materials*, 4, 435–446.
5. Murphy, C. J., Gole, A. M., Stone, J. W., Sisco, P. N., Alkilany, A. M., Goldsmith, E. C. and Baxter, S. C. 2008. Gold nanoparticles in biology: Beyond toxicity to cellular imaging. *Accounts of Chemical Research*, 41, 1721–1730.
6. Jana, N. R. 2011. Design and development of quantum dots and other nanoparticles based cellular imaging probe. *Physical Chemistry Chemical Physics*, 13, 385–396.
7. Zrazhevskiy, P., Sena, M. and Gao, X. H. 2010. Designing multifunctional quantum dots for bioimaging, detection, and drug delivery. *Chemical Society Reviews*, 39, 4326–4354.
8. Sapsford, K. E., Algar, W. R., Berti, L., Gemmill, K. B., Casey, B. J., Oh, E., Stewart, M. H. and Medintz, I. L. 2013. Functionalizing nanoparticles with biological molecules: Developing chemistries that facilitate nanotechnology. *Chemical Reviews*, 113, 1904–2074.
9. Moyano, D. F. and Rotello, V. M. 2011. Nano meets biology: Structure and function at the nanoparticle interface. *Langmuir*, 27, 10376–10385.
10. Barua, S. and Mitragotri, S. 2014. Challenges associated with penetration of nanoparticles across cell and tissue barriers: A review of current status and future prospects. *Nano Today*, 9, 223–243.

11. Martens, T. F., Remaut, K., Demeester, J., De Smedt, S. C. and Braeckmans, K. 2014. Intracellular delivery of nanomaterials: How to catch endosomal escape in the act. *Nano Today*, 9, 344–364.
12. Fleischer, C. C. and Payne, C. K. 2014. Nanoparticle-cell interactions: Molecular structure of the protein corona and cellular outcomes. *Accounts of Chemical Research*, 47, 2651–2659.
13. Ding, H. M. and Ma, Y. Q. 2015. Theoretical and computational investigations of nanoparticle-biomembrane interactions in cellular delivery. *Small*, 11, 1055–1071.
14. Mornet, S., Vasseur, S., Grasset, F. and Duguet, E. 2004. Magnetic nanoparticle design for medical diagnosis and therapy. *Journal of Materials Chemistry*, 14, 2161–2175.
15. Basiruddin, S. K., Saha, A., Pradhan, N. and Jana, N. R. 2010. Advances in coating chemistry in deriving soluble functional nanoparticle. *The Journal of Physical Chemistry C*, 114, 11009–11017.
16. Erathodiyil, N. and Ying, J. Y. 2011. Functionalization of inorganic nanoparticles for bioimaging applications. *Accounts of Chemical Research*, 44, 925–935.
17. Ling, D. S., Hackett, M. J. and Hyeon, T. 2014. Surface ligands in synthesis, modification, assembly and biomedical applications of nanoparticles. *Nano Today*, 9, 457–477.
18. Ding, C. Q., Zhu, A. W. and Tian, Y. 2014. Functional surface engineering of C-dots for fluorescent biosensing and *in vivo* bioimaging. *Accounts of Chemical Research*, 47, 20–30.
19. Fratila, R. M., Mitchell, S. G., del Pino, P., Grazu, V. and de la Fuente, J. M. 2014. Strategies for the biofunctionalization of gold and iron oxide nanoparticles. *Langmuir*, 30, 15057–15071.
20. Chen, Y. P., Xianyu, Y. L. and Jiang, X. Y. 2017. Surface modification of gold nanoparticles with small molecules for biochemical analysis. *Accounts of Chemical Research*, 50, 310–319.
21. Chakraborty, A., Dalal, C. and Jana, N. R. 2018. Colloidal nanobioconjugate with complementary surface chemistry for cellular and subcellular targeting. *Langmuir*, 34, 13461–13471.
22. Prasad, B. L. V., Stoeva, S. I., Sorensen, C. M. and Klabunde, K. J. 2002. Digestive ripening of thiolated gold nanoparticles: The effect of alkyl chain length. *Langmuir*, 18, 7515–7520.
23. Li, J. J., Wang, Y. A., Guo, W. Z., Keay, J. C., Mishima, T. D., Johnson, M. B. and Peng, X. G. 2003. Large-scale synthesis of nearly monodisperse CdSe/CdS core/shell nanocrystals using air-stable reagents via successive ion layer adsorption and reaction. *Journal of the American Chemical Society*, 125, 12567–12575.

24. Jana, N. R., Chen, Y. and Peng, X. 2004. Size- and shape-controlled magnetic (Cr, Mn, Fe, Co, Ni) oxide nanocrystals via a simple and general approach. *Chemistry of Materials*, 16, 3931–3935.
25. Zhang, Q., Xie, J., Yang, J. and Lee, J. Y. 2009. Monodisperse icosahedral Ag, Au and Pd nanoparticles: Size control strategy and superlattice formation. *ACS Nano*, 3, 139–148.
26. Srivastava, B. B., Jana, S., Karan, N. S., Paria, S., Jana, N. R., Sarma, D. D. and Pradhan, N. 2010. Highly luminescent Mn-doped ZnS nanocrystals: Gram scale synthesis. *The Journal of Physical Chemistry Letters*, 1, 1454–1458.
27. Jana, N. R., Erathodiyil, N., Jiang, J. and Ying, J. Y. 2010. Cyeteine-functionalized polyaspartice acid: A polymer for coating and bioconjugation of nanoparticles and quantum dots. *Langmuir*, 26, 6503–6507.
28. Wei, H., Insin, N., Lee, J., Han, H.-S., Cordero, J. M., Liu, W. and Bawendi, M. W. 2012. Compact zwitterion-coated iron oxide nanoparticles for biological applications. *Nano Letters*, 12, 22–25.
29. Zhan, N., Palui, G., Safi, M., Ji, X. and Mattoussi, H. 2013. Multidentate zwitterionic ligands provide compact and highly biocompatible quantum dots. *Journal of the American Chemical Society*, 135, 13786–13795.
30. Wang, W., Kapur, A., Ji, X., Zeng, B., Mishra, D. and Mattoussi, H. 2016. Multifunctional and high affinity polymer ligand that provides bio-orthogonal coating of quantum dots. *Bioconjugate Chemistry*, 27, 2024–2036.

CHAPTER 2

Design Considerations of Nanoparticle Functionalization

2.1 INTRODUCTION

A variety of nanoparticles are known, along with their established synthetic methods. The physical and chemical properties of nanoparticles vary depending on size, shape, composition, surface chemistry, colloidal properties and surface functional groups. Thus, appropriate selection of nanoparticles is required for the intended application. More importantly, interaction with a biological interface depends on the physical and chemical properties of nanoparticles. This chapter focuses on the design considerations for nanoparticles that are required for their biomedical application. This chapter offers general guidelines about the basic requirements and why such requirements are necessary.

2.2 SIZE

The appropriate size of nanoparticles is important in obtaining their physical/chemical properties and for their application in biomedical science.[1-8] As most nanoparticles have size-dependent properties, the appropriate size is required to utilize such properties. For example, gold nanoparticle size needs to be controlled in the 5–100 nm range to exploit their plasmonic properties, and the size of CdSe-based quantum dots needs to be controlled in the 2–6 nm range to utilize their size-dependent emission properties.[1-4]

Moreover, from a biomedical application point of view, nanoparticles are unique, as their size is similar to many biomolecules or biological units, and nanoparticles can read biological functions/ activity[3,4,9-12] (Figure 2.1). For example, the optical property of nanoparticles can be used to monitor protein–protein interactions, protein-DNA interactions, the adsorption properties of biomolecules on solid surfaces, cellular interactions of proteins and cellular endocytosis.[3,4,9-12] Similarly, nanoscale drug delivery carriers can be used for targeting/delivery at specific organ, cell or subcellular targets, and nanoscale imaging agents can be used to monitor the

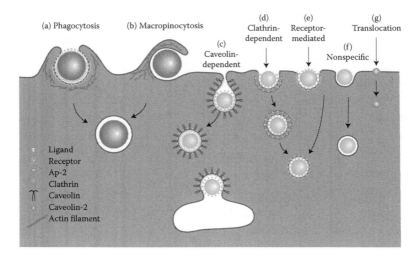

FIGURE 2.1 Effect of nanoparticle size and shape in cellular endocytosis (a–e), non-specific uptake (f), and direct translocation (g). (Reprinted with permission from Zhang, S. et al. Physical principles of nanoparticle cellular endocytosis. *ACS Nano*, 9, 8655–8671. Copyright 2015 American Chemical Society.)

biochemical activity or progress of a disease at the subcellular length scale.[10] It has been shown that nanoparticles with a hydrodynamic size of <5.5 nm are ideal for renal clearance, 10–50 nm is ideal for cellular and subcellular targeting and 50–200 nm is ideal for enhanced permeation retention (EPR)-based *in vivo* targeting applications.[10] Thus, design/selection of the appropriate size is essential for the intended biomedical application of nanoparticles.

2.3 SHAPE

The shape of nanoparticles can dictate their physical/chemical properties and biological performance.[1,2,5,11–15] A unique example is a gold nanorod that has a length-dependent longitudinal plasmon band extending from the visible to the near-infrared (NIR) region, in addition to the transverse plasmon band at 500 nm that commonly appears for spherical gold nanoparticles. Similarly, gold nanowire, silver nanorods, silver nanowire, CdSe nanorods, ZnO nanorods, cubic gold nanoparticles and silver nanoplates have unique physical and chemical properties.[1,2,5,13,14]

The interaction of nanoparticles with a biological interface can also depend on particle shape[12,15] (Figure 2.1). For example, cellular interaction and endocytosis of nanoparticles depends on nanoparticle shape. It has been shown that anisotropic nanoparticles have higher cell uptake than similarly sized spherical nanoparticles, and the mechanism of cellular entry depends on the length-to-width ratio of the nanoparticle.[12,15] Similarly, wire like nanoparticles have a better *in vivo* targeting property than similarly sized spherical nanoparticles, as nanowires can stay longer in blood by aligning along blood flow.[16] Thus, designing an appropriate shape of nanoparticle may be critical for some specific biomedical applications.

2.4 COMPOSITION AND PHYSICAL PROPERTIES

Nanoparticles are varied in their chemical composition and physical forms. For example, inorganic nanoparticles are prepared from metal, metal oxide, metal sulfide and metal selenide.[10,11] Similarly, various organic molecules and polymers are used to make organic nanoparticles.[17] In most cases, nanoparticles are in solid powder form or in a colloidal state. Thus, the chemical composition and physical forms of nanoparticles play a critical role toward biomedical application, particularly because the physical/chemical properties of nanoparticles depend on their chemical composition. For example, the chemical compositions of nanoparticles dictate their optical and magnetic properties and colloidal state direct their interaction/reactivity to biological interfaces. As the optical/magnetic properties will be utilized for specific biomedical applications, chemical compositions need to be appropriately selected. Similarly, chemical compositions need to be appropriately selected so that they have minimum reaction and toxicity to biological systems.

2.5 BIOCOMPATIBILITY AND BIODEGRADABILITY

The most critical criteria for biomedical applications are that nanoparticles be biocompatible, environment friendly and preferably biodegradable. Biocompatible nanoparticles would ensure intact biological function during their application. In addition, toxic nanoparticles will have limited application potential to living systems and an adverse effect on the environment, particularly in their large-scale use. Thus, an attempt should be made to prepare nanoparticles from nontoxic chemicals and preferably from biological or biocompatible chemicals.

2.6 HYBRID AND MULTIFUNCTIONAL NANOPARTICLES

In many applications, hybrid nanoparticles with multifunctional properties are used, with specific advantages.[18] For example, magnetic-fluorescent nanoparticles are prepared by using a composite between magnetic and fluorescent nanoparticles (or fluorescent dye), and these nanoparticles are used for simultaneous optical detection and magnetic separation (Figure 2.2). Similarly, plasmonic-fluorescent nanoparticles are used as optical probes with a detection option via both fluorescence and dark field.

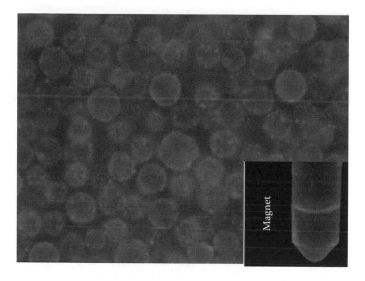

FIGURE 2.2 Fluorescence image of cell labeled with magnetic-fluorescent nanoparticle and magnetic separation of labeled cell by laboratory-based bar magnet. (Reprinted with permission from Saha, A, et al. Ligand exchange approach in deriving magnetic-fluorescent and magnetic–plasmonic hybrid nanoparticle. *Langmuir*, 26, 4351–4356 Copyright 2010 American Chemical Society.)

Fluorescent nanocarriers are designed with the option for delivery and tracking of drugs inside cells. These examples show that hybrid nanoparticles have unique advantages and, depending on the application requirements, can be designed.

2.7 WATER SOLUBILITY AND COLLOIDAL STABILITY

Biomedical application requires that nanoparticles be in their colloidal form in water during their interaction with a biological interface.[7-11] In the colloidal form, they are easily accessible by remote targets such as organs, cells and subcellular compartments. However, designing a colloidal form of a nanoparticle that is stable in a complex biological environment is extremely critical. This is because colloids are stable due to surface charge, but the physiological environment is associated with a reasonably high salt concentration that screens the surface charge. In addition, charged polymers (e.g., proteins/DNA) are present at high concentrations that are adsorbed on the particle surface via electrostatic interaction. These factors often decrease the colloidal stability of nanoparticles/ nanobioconjugates during their biomedical application. Thus, attempts should be made to design nanoparticles that are stable under physiological conditions. In particular, nanoparticles should be grafted with nonionic polymers to induce steric stabilization or with zwitterionic polymers to minimize electrostatic interaction with foreign biomolecules.

2.8 SURFACE CHEMISTRY, CHARGE AND LIPOPHILICITY

The surface chemistry of nanoparticles is the most critical property when they are interacting with a biological interface.[19] In particular, the nature of capping molecules dictates their hydrophobic/hydrophilic properties. Hydrophobic surfactant-capped nanoparticles are water insoluble and dispersible in organic solvent. Such hydrophobic nanoparticles include dodecanethiol-capped Au nanoparticles, octadecylamine-capped iron oxide nanoparticles, oleylamine-capped Au nanoparticles and hexadecylamine-capped quantum dots. In contrast, hydrophilic molecule/surfactant/polymer-capped nanoparticles are water soluble. Such hydrophilic nanoparticles include citrate-capped Au nanoparticles, cetyltrimethylammonium bromide-capped Au nanorods and mercaptoundecanoic acid-capped Au nanoparticles.

The colloidal form of nanoparticles is usually charged, either cationic or anionic. Anionic nanoparticles are more common in nature because of the adsorption of anions on their surface. However, nanoparticles can be designed with controlled cationic/ anionic charge. In particular, positively charged nanoparticles can be prepared by using a cationic surfactant or polymer as a nanoparticle stabilizer, and anionic nanoparticles can be prepared by using an anionic surfactant/polymer as a stabilizer. For example, cetyltrimethylammonium bromide-capped nanoparticles show

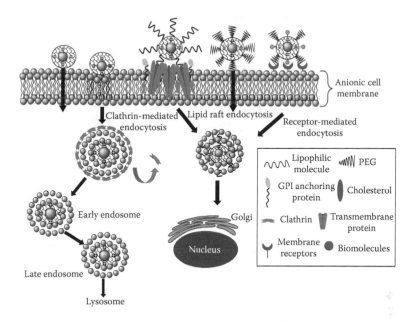

FIGURE 2.3 Surface chemistry dependent cellular interaction and endocytosis of nanoparticles. PEG and GPI refer to polyethylene glycol and glycophosphatidylinositol, respectively. (Reprinted with permission from Chakraborty, A., Dalal, C. and Jana, N. R. 2018. Colloidal nanobioconjugate with complementary surface chemistry for cellular and subcellular targeting. *Langmuir*, 34, 13461–13471. Copyright 2016 American Chemical Society.)

positive zeta potential, and sodium dodecyl sulfate-capped nanoparticles show negative zeta potential.

The interaction of nanoparticles with live cells is sensitive to nanoparticle surface charge and lipophilicity (Figure 2.3). For example, cationic nanoparticles strongly interact with anionic cell membranes via electrostatic interaction, and lipophilic nanoparticles can also strongly interact with lipidic cell membranes. In contrast, anionic nanoparticles do not interact with cell membranes. Similarly, a high cationic surface charge induces cell membrane damage and cytotoxicity. Thus, appropriate design of a nanoparticle surface is key for the desired biomedical application (see Figure 2.3).

2.9 BIOFUNCTIONALITY

As nanoparticles interact with a biological interface through the functional groups at their surface, it is important to attach appropriate biomolecules at their surface so that nanoparticles can selectively interact with the biological interface.[11,20-24] Commonly, various biomolecules are covalently attached on their surface. These biomolecules include vitamins, antibodies, peptides, oligonucleotides, aptamers, carbohydrates, glycoproteins and so on.[11,20-24] Attachment of these biomolecules on the nanoparticle surface would offer their selective interaction/labeling with the biological interface. Thus, the relevant surface and conjugation chemistry need to be adapted to transform as-synthesized nanoparticles into designed nanobioconjugates.

2.10 NONSPECIFIC INTERACTION

As nanoparticles have high surface area and are decorated with multiple chemical functional groups, they often induce nonspecific interaction with a biological interface. In particular, nanoparticles often have stronger interaction with proteins that changes their surface properties. Nanoparticles strongly interact with cell membranes that enhance their cellular endocytosis.[20,25] These factors often inhibit their specific labeling application. In a common approach, a nanoparticle surface is covered with polyethylene glycol, dextran or other nonionic polymers to minimize such nonspecific interactions, and these approaches should be utilized in designing nanobioconjugates.[20]

2.11 CONTROLLING NANOPARTICLE MULTIVALENCY

Most nanobioconjugates are essentially multivalent in nature. This is because nanoparticles have multiple biomolecules on their surface and interact with the biological interface via these biomolecules.[26-28] Conventional nanobioconjugates typically have a few hundred to a few thousand biomolecules

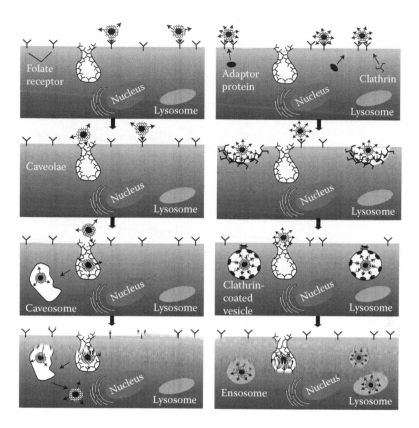

FIGURE 2.4 Multivalency of nanoparticles can be used to control their entry mechanism into live cells. (Reprinted with permission from Dalal, C. et al. Nanoparticle multivalency directed shifting of cellular uptake mechanism. *The Journal of Physical Chemistry C*, 120, 6778–6786. Copyright 2016 American Chemical Society.)

per nanoparticle on their surface, meaning that multivalency of nanobioconjugates is in a similar range. It has been shown that this multivalency dictates the interaction of nanoparticles with DNA/proteins/cells and controls biological interaction and the cellular endocytosis mechanism (Figure 2.4). Thus, an attempt should be made to control this multivalency for effective use of nanoparticles.

REFERENCES

1. El-Sayed, M. A. 2004. Small is different: Shape-, size-, and composition-dependent properties of some colloidal semiconductor nanocrystals. *Accounts of Chemical Research*, 37, 326–333.
2. Sau, T. K. and Murphy, C. J. 2004. Room temperature, high-yield synthesis of multiple shapes of gold nanoparticles in aqueous solution. *Journal of the American Chemical Society*, 126, 8648–8649.
3. Michalet, X., Pinaud, F. F., Bentolila, L. A., Tsay, J. M., Doose, S., Li, J. J., Sundaresan, G., Wu, A. M., Gambhir, S. S. and Weiss, S. 2005. Quantum dots for live cells, *in vivo* imaging, and diagnostics. *Science*, 307, 538–544.
4. Medintz, I. L., Uyeda, H. T., Goldman, E. R. and Mattoussi, H. 2005. Quantum dot bioconjugates for imaging, labelling and sensing. *Nature Materials*, 4, 435–446.
5. Jain, P. K., Huang, X. H., El-Sayed, I. H. and El-Sayed, M. A. 2008. Noble metals on the nanoscale: Optical and photothermal properties and some applications in imaging, sensing, biology, and medicine. *Accounts of Chemical Research*, 41, 1578–1586.
6. Murphy, C. J., Gole, A. M., Stone, J. W., Sisco, P. N., Alkilany, A. M., Goldsmith, E. C. and Baxter, S. C. 2008. Gold nanoparticles in biology: Beyond toxicity to cellular imaging. *Accounts of Chemical Research*, 41, 1721–1730.
7. Jana, N. R. 2011. Design and development of quantum dots and other nanoparticles based cellular imaging probe. *Physical Chemistry Chemical Physics*, 13, 385–396.
8. Cao, H., Ma, J., Huang, L., Qin, H., Meng, R., Li, Y. and Peng, X. 2016. Design and synthesis of antiblinking and antibleaching quantum dots in multiple colors via wave functional confinement. *Journal of the American Chemical Society*, 138, 15727–15735.
9. Mahtab, R., Jessica, P., Rogers, J. P. and Murphy, C. J. 1995. Protein-sized quantum dot luminescence can distinguish between "straight," "bent," and "kinked" oligonucleotides. *Journal of the American Chemical Society*, 117, 9099–9100.
10. Moyano, D. F. and Rotello, V. M. 2011. Nano meets biology: Structure and function at the nanoparticle interface. *Langmuir*, 27, 10376–10385.
11. Sapsford, K. E., Algar, W. R., Berti, L., Gemmill, K. B., Casey, B. J., Oh, E., Stewart, M. H. and Medintz, I. L. 2013. Functionalizing nanoparticles with biological molecules: Developing chemistries that facilitate nanotechnology. *Chemical Reviews*, 113, 1904–2074.

12. Zhang, S., Gao, H. and Bao, G. 2015. Physical principles of nanoparticle cellular endocytosis. *ACS Nano*, 9, 8655–8671.
13. Lohse, S. E. and Murphy, C. J. 2013. The quest for shape control: A history of gold nanorod synthesis. *Chemistry of Materials*, 25, 1250–1261.
14. Yang, J., Son, J. S., Yu, J. H., Joo, J. and Hyeon, H. 2013. Advances in the colloidal synthesis of two-dimensional semiconductor nanoribbons. *Chemistry of Materials*, 25, 1190–1198.
15. Dasgupta, S., Auth, T. and Gompper, G. 2014. Shape and orientation matter for the cellular uptake of nanospherical particles. *Nano Letters*, 14, 687–693.
16. Nishiyama, N. 2007. Nanocarriers shape up for long life. *Nature Nanotechnology*, 2, 203–204.
17. Haladjova, E., Toncheva-Moncheva, N., Apostolova, M. D., Trzebicka, B., Dworak, A., Petrov, P., Dimitrov, I., Rangelov, S. and Tsvetanov, C. B. 2014. Polymer nanoparticle engineering: From temperature responsive polymer mesoglobules to gene delivery systems. *Biomacromolecules*, 15, 4377–4395.
18. Saha, A., Basiruddin, S. K., Pradhan, N. and Jana, N. R. 2010. Ligand exchange approach in deriving magnetic–fluorescent and magnetic–plasmonic hybrid nanoparticle. *Langmuir*, 26, 4351–4356.
19. Basiruddin, S. K., Saha, A., Pradhan, N. and Jana, N. R. 2010. Advances in coating chemistry in deriving soluble functional nanoparticle. *The Journal of Physical Chemistry C*, 114, 11009–11017.
20. Chakraborty, A., Dalal, C. and Jana, N. R. 2018. Colloidal nanobioconjugate with complementary surface chemistry for cellular and subcellular targeting. *Langmuir*, 34, 13461–13471.
21. Erathodiyil, N. and Ying, J. Y. 2011. Functionalization of inorganic nanoparticles for bioimaging applications. *Accounts of Chemical Research*, 44, 925–935.
22. Ling, D. S., Hackett, M. J. and Hyeon, T. 2014. Surface ligands in synthesis, modification, assembly and biomedical applications of nanoparticles. *Nano Today*, 9, 457–477.
23. Ding, C. Q., Zhu, A. W. and Tian, Y. 2014. Functional surface engineering of C-dots for fluorescent biosensing and *in vivo* bioimaging. *Accounts of Chemical Research*, 47, 20–30.
24. Fratila, R. M., Mitchell, S. G., del Pino, P., Grazu, V. and de la Fuente, J. M. 2014. Strategies for the biofunctionalization of gold and iron oxide nanoparticles. *Langmuir*, 30, 15057–15071.

25. Barua, S. and Mitragotri, S. 2014. Challenges associated with penetration of nanoparticles across cell and tissue barriers: A review of current status and future prospects. *Nano Today*, 9, 223–243.
26. Saha, A., Basiruddin, S. K., Maity, A. R. and Jana, N. R. 2013. Synthesis of nanobioconjugates with a controlled average number of biomolecules between 1 and 100 per nanoparticle and observation of multivalency dependent interaction with proteins and cells. *Langmuir*, 29, 13917–13924.
27. Dalal, C., Saha, A. and Jana, N. R. 2016. Nanoparticle multivalency directed shifting of cellular uptake mechanism. *The Journal of Physical Chemistry C*, 120, 6778–6786.
28. Dalal, C. and Jana, N. R. 2017. Multivalency effect of TAT-peptide-functionalized nanoparticle in cellular endocytosis and subcellular trafficking. *The Journal of Physical Chemistry B*, 121, 2942–2951.

CHAPTER 3

Chemical Synthetic Methods of Selected Nanoparticles

3.1 INTRODUCTION

Synthesis of nanoparticles by a simple and reliable approach is most critical for their biomedical application. This chapter focuses on standard synthetic methods of some selected nanoparticles. Nanoparticles are selected based on their wider application potential and routine use in various scientific disciplines. Synthetic methods are selected that are simple, reliable and most frequently used and have good reproducibility. Nanoparticles include two types of plasmonic materials of different sizes (Au and Ag), one magnetic material with different sizes (iron oxide), one QD material with different emission colors (ZnS-capped CdSe), fluorescent carbon nanoparticles, two doped semiconductor nanoparticles (Mn-doped ZnS and N, F codoped TiO_2), two oxide nanoparticles (ZnO, TiO_2), hydroxyapatite nanorods, graphene oxide/chemically reduced graphene oxide, liposome, albumin nanoparticles and two polymer nanoparticles. The described synthetic methods can assist graduate students and researchers with the exposure to the best synthetic methods from the vast body of literature. This chapter will also offer them general guidelines and the confidence to prepare other new nanoparticles of similar types following the appropriate literature.

3.2 Au NANOPARTICLES

3.2.1 Citrate-Capped Au (3–4 nm)

Take 20 mL aqueous solution of $HAuCl_4$ (0.25 mM) in a conical flask.[1] In a separate vial, prepare an aqueous solution of trisodium citrate with a final concentration of 2.5 mM and add 2.0 mL of it to a gold salt solution. Keep the solution under stirring. Next, add 0.6 mL of freshly prepared 0.1 M $NaBH_4$ solution and stop stirring after 2–3 min. The solution turns pink immediately after adding $NaBH_4$. The particles are generally stable for 2–5 h after preparation.

3.2.2 Citrate-Capped Au (12–20 nm)

Take 10 mL of aqueous $HAuCl_4$ (0.25 mM) in a glass vial and heat it to boiling with the continuous stirring condition.[2] Next, add 3 mL of 1% aqueous trisodium citrate. The solution undergoes color changes from blue to wine red. Continue the boiling and stirring for 30 min and cool to room temperature. This method produces Au nanoparticles with a mean diameter of 12–20 nm.

3.2.3 Surfactant-Capped Hydrophilic Au Nanoparticles via Seeding Growth (5–50 nm)

Use citrate-capped 3–4-nm Au nanoparticles as a seed, as described in Section 3.2.1.[3] Prepare 200 mL aqueous solution of $HAuCl_4$ (2.5×10^{-4} M) in a conical flask. Next, add 6 g of solid cetyltrimethylammonium bromide (CTAB) to the solution and heat to 50°C–60°C until the solution turns a clear orange color. Cool the solution to room temperature and use as a stock growth solution.

Next, take four sets of 50 mL conical flasks labeled A, B, C and D. In set A, add 7.5 mL of growth solution and mix with 0.05 mL of freshly prepared ascorbic acid (0.1 M) solution. Next, add 2.5 mL of seed solution (all at once) while stirring. Continue stirring for

10 min until the solution turns wine red. Particles prepared this way are spherical with a diameter of 5–6 nm.

In set B, mix 9 mL of growth solution with 0.05 mL ascorbic acid (0.1 M), then mix with 1.0 mL of seed solution. The final color of the solution becomes a deep red within 15–30 min. Particles prepared this way are spherical with a diameter of 8.0 nm.

In set C, mix 9 mL of growth solution with 0.05 mL ascorbic acid (0.1 M) solution, then mix with 1.0 mL solution from set B. The final color of the solution becomes reddish brown within 1 h. Particles prepared in this way are roughly spherical with a diameter of 15–20 nm.

In set D, mix 9 mL of growth solution with 0.05 mL of ascorbic acid (0.1 M) solution, then mix with 1.0 mL solution from set C. The final color of the solution becomes brown within 1–2 h. Particles prepared this way have a 35–40-nm diameter. Solutions A, B, C and D are stable for more than a month due to the presence of CTAB as a particle stabilizer. The gold concentration in A–D is 2.5×10^{-4} M. Excess CTAB can be removed by precipitating the particles by high-speed centrifuging and redispersing the particles in fresh water. A TEM image of Au nanoparticles prepared via the seeding growth method is shown in Figure 3.1.

3.2.4 Weakly Adsorbed Surfactant-Capped Hydrophobic Au Nanoparticles (2–5 nm)

Prepare a toluene solution of 5 mL $AuCl_3$ (0.01 M) in the presence of didodecyldimethylammonium bromide (DDAB, 0.02 M).[4] Next, add 100 µL oleyl amine and 100 µL oleic acid to this solution. In a separate vial, dissolve 25 mg tetrabutylammonium borohydride (TBAB) in the presence of 25 mg DDAB in 500 µL toluene and inject the whole solution into the earlier gold salt solution. A deep brown color appears immediately after adding TBAB. Precipitate the synthesized nanoparticles by adding the minimum ethanol, followed by centrifugation to remove excess reagent. Finally, dissolve the precipitated nanoparticles in 5 mL cyclohexane and use as stock solution.

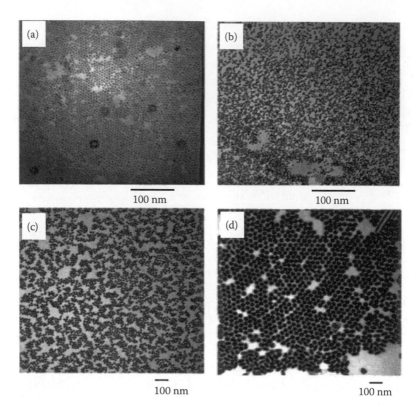

FIGURE 3.1   TEM image of different-sized Au nanoparticles prepared via seeding growth method. (a) 5.5 nm, (b) 8 nm, (c) 17 nm and (d) 37 nm. (Reprinted with permission from Jana, N. R. et al. Seeding growth for size control of 5–40 nm diameter gold nanoparticles. *Langmuir*, 17, 6782–6786. Copyright 2001 American Chemical Society.)

3.2.5 Thiol-Capped Hydrophobic Au (2–6 nm)

Dissolve 7.5 mg $AuCl_3$ in 2.5 mL toluene solution of didodecylammonium bromide (0.1M).[4–7] Add 0.1–0.5 mL toluene solution of dodecanethiol solution (1 M). The solution turns faint yellow within 1–2 min of adding thiol. In a separate vial, dissolve 25 mg tetrabutyl ammonium borohydride and 25 mg DDAB in 1 mL toluene and inject the whole solution into the gold salt solution during stirring. A dark brown color appears immediately that becomes stable for more than 24 h. Next, heat the solution to 80°C–100°C for 5 min to 2 h. The size of the Au nanoparticles will increase from 2 to 6 nm as the heating time increases. Nanoparticles can be precipitated by adding ethanol and redispersed in fresh toluene/chloroform.

3.3 Au NANORODS

3.3.1 Au Nanorods via Seedless Approach
(5–10 nm × 20–50 nm)

Prepare 10 mL of aqueous solution of CTAB surfactant containing $HAuCl_4$, where the final concentrations of CTAB and $HAuCl_4$ are 0.1 M and 1 mM, respectively.[8] Next, add 0.1 mL aqueous solution of $AgNO_3$ (0.02 M). Next, add 0.1 mL aqueous solution of ascorbic acid (0.2 M), which changes the solution color from orange to colorless. Finally, add 10–100 μL aqueous solution of freshly prepared sodium borohydride (1 mM). A pink-violet solution will appear within 1 h, indicating the formation of gold nanorods. Ideally, prepare five sets using varied volumes of borohydride (10, 20, 30, 50 and 100 μL) and select the one or two with clear nanorod plasmon bands (strong longitudinal plasmon band but weak transverse band) for further use. Remove the excess CTAB by precipitating nanorods via high-speed centrifuge (at 12,000–15,000 rpm) and redissolving nanorods in fresh water. This precipitation-redispersion can be repeated two to three times to remove most of the free CTAB. TEM image of Au nanorods with controlled length-to-width ratio prepared by this approach is shown in Figure 3.2.

3.3.2 Au Nanorods via Seeding Growth
Approach (10–20 nm × 20–100 nm)

First, prepare Au seeds of 1.5 nm size by rapidly injecting $NaBH_4$ (200 mL, 0.1 M) solution into a vigorously stirred 10 mL aqueous solution containing $HAuCl_4$ (0.25 mM), CTAB (0.1 M) and tetraethyl ammonium hydroxide (0.01 M).[9–11] Stop stirring after 2–3 min and use this as seed solution after 2 h of preparation when

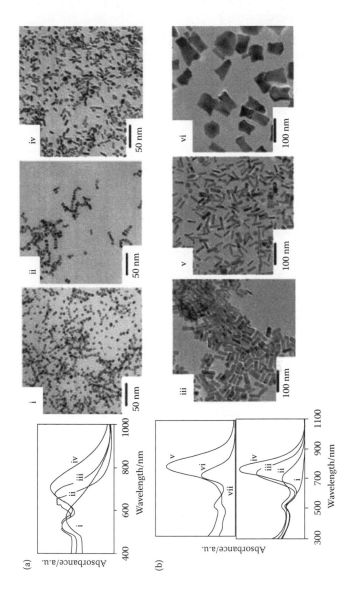

FIGURE 3.2 TEM image and optical property of Au nanorods prepared via seedless approach where (a[i, ii, iv]) and (b[iii, v, vi]) represent different nucleation-growth conditions. (Jana, N. R. 2005. Gram-scale synthesis of soluble, nearmonodisperse gold nanorods and other anisotropic nanoparticles. *Small*, 1, 875–882. Copyright Wiley-VCH Verlag GmbH & Co. KGaA. Reproduced with permission.)

no excess borohydrides are present. It can be used as stock seed solution for the next 2–3 days.

Next, prepare 10 mL growth solution containing $HAuCl_4$ (1.0 mM), $AgNO_3$ (0.2 mM) and CTAB (0.2 M). Then, add 0.2 mL ascorbic acid (0.1 M) to the growth solution, which leads to a change of the orange-colored solution to colorless. Next, mix with 0.01–0.1 mL of seed solution and wait for 2–3 h. Within 30–60 min, the solution color changes to brown/violet/ green, depending on the volume of seed added. Ideally, prepare five sets using varied volumes of seed solution (10, 20, 30, 50 and 100 μL) and select the one or two with clear nanorod plasmon bands (strong longitudinal plasmon band and weak transverse band) for further use. Remove the excess CTAB by high-speed centrifuge (12,000–15,000 rpm) that precipitates nanorods, and redissolve the nanorods in fresh water. This precipitation-redispersion can be repeated two or three times to remove most of the free CTAB. Figure 3.3 shows a typical TEM image of Au nanorods prepared via the seeding growth approach.

3.4 Ag NANOPARTICLES

3.4.1 Citrate-Capped Ag (20–30 nm)

Prepare 10 mL $AgNO_3$ solution (1 mM) in a conical flask and heat it to the boiling condition.[12] Next, add 200 μL of 1% sodium citrate solution into the boiling solution and continue boiling for 90 min with the resultant deep yellow solution. You can add water (vaporated) during or after the reaction to make up the volume. Preserve the solution as stock, as the colloidal solution is stable for months. If you see some partial precipitation, shake it well before use.

3.4.2 Surfactant- and Thiol-Capped Hydrophobic Ag Nanoparticles (2–5 nm)

Dissolve 17 mg Ag-acetate in 9 mL toluene in the presence of 260 μL of octylamine. Next, add 100 μL of oleic acid to the same solution.[4] In a separate vial, dissolve 50 mg tetrabutyl ammonium borohydride in a mixture of 1 mL toluene and 100 μL oleic acid.

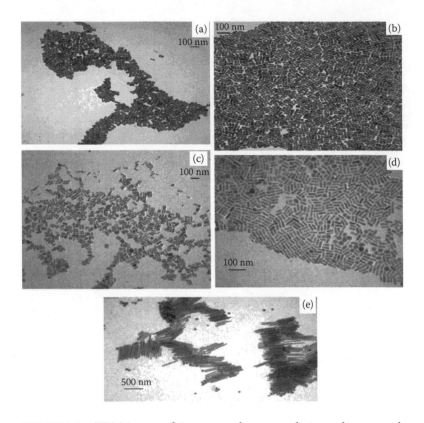

FIGURE 3.3 TEM image of Au nanorods prepared via seeding growth approach where (a)–(e) represent different seeding growth condition. (Reprinted with permission Sau, T. K. and Murphy, C. J. 2004. Seeded high yield synthesis of short Au nanorods in aqueous solution. *Langmuir*, 20, 6414–6420. Copyright 2004 American Chemical Society.)

(CAUTION! Hydrogen gas evolution occurs upon oleic acid addition.) Inject the whole solution into the silver salt solution. A yellow solution appears immediately after adding borohydride. Next, add the minimum anhydrous ethanol (10–20 mL) to precipitate the particles. Isolate the particles by centrifuging at 6000 rpm for 3–4 min and dissolve in 10 mL of cyclohexane for further use. The same reaction can be performed at a small scale using 1/10 of all the reagents and solvents.

Dodecanetiol-capped Ag nanoparticles of 2–5 nm can be prepared by adding two to five equivalent (0.5–0.01 mL) dodecanethiol before adding borohydride, followed by heating at 90°C–100°C for up to 2 h.

3.5 Ag-CAPPED Au (Ag@Au) NANOPARTICLES (25–30 nm)

Monodispersed Ag nanoparticles are relatively difficult to prepare in comparison to monodispersed Au nanoparticles.[13] However, Ag-coated Au nanoparticles (Ag@Au) can be prepared from monodispersed Au nanoparticles via the seeding growth approach. The resultant Ag@Au are monodispersed and have dominating Ag plasmons (Figure 3.4). Here is one procedure that involves deposition of Ag on the surface of 12-nm Au nanoparticles. Prepare 12 nm colloidal Au nanoparticles by the citrate reduction method described in Section 3.2.2. Next, mix 10 mL aqueous solution of Au nanoparticles with 50 mL ascorbic acid (0.1 M) solution. Next, add 0.25 mL $AgNO_3$ (0.01 M) solution in a dropwise manner within 5 min under the continuous stirring condition. The solution color changes from deep yellow to gray. The resultant Ag@Au is colloidally stable for several weeks and should be used within this period.

3.6 HYDROPHOBIC CdSe-ZnS QUANTUM DOTS (2–6 nm)

Take 23 mg CdO, 300 mg stearic acid and 10 mL octadecene (ODE) in a 25-mL three-necked round-bottomed flask and keep under the magnetic stirring condition.[14,15] Degas the solution with argon for 15 minutes below 100°C, then heat the solution to about 220°C to obtain a colorless solution of $Cd(stearate)_2$. Then, cool the solution to room temperature and add 1.5 g octadecylamine (ODA) and 0.5 g trioctylphosphine oxide (TOPO) into the reaction flask. Under argon flow, raise the temperature of the reaction mixture to 280°C. In a separate vial, take 40 mg of selenium powder in 1 mL trioctylphosphine (TOP) and dissolve at 80°C–100°C, then inject

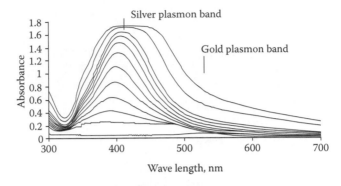

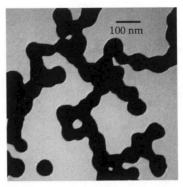

FIGURE 3.4 Optical property and TEM image of Ag-capped Au nanoparticles prepared using 12 nm Au seed. (Jana, N. R. 2003. Silver coated gold nanoparticles as new surface enhanced Raman substrate at low analyte concentration. *Analyst*, 128, 954–956. Reproduced by permission of The Royal Society of Chemistry.)

the selenium solution quickly. Maintain the growth temperature for 15 sec to 15 min, then rapidly cool to room temperature. Typically, this reaction generates CdSe nanocrystals of about 2–6 nm in size with blue/green/yellow/red emission. In order to purify CdSe nanocrystals from unreacted precursors, add acetone such that the volume ratio of ODE and acetone is 1:1. Then, centrifuge to precipitate the CdSe nanocrystal and redisperse in the minimum amount of chloroform. Next, add the minimum ethanol to the chloroform solution of CdSe nanocrystals to precipitate the CdSe nanocrystals. Collect the precipitate by centrifugation and again redisperse in 5 mL ODE for ZnS shelling on the CdSe nanocrystals. Typically, take the 5 mL ODE solution of the CdSe nanocrystals in a 25-mL three-necked flask. Then, add 2.8 g ODA and 1 g TOPO to the CdSe solution and degas the reaction mixture under argon flow with 60°C–80°C heating conditions for 20 min. Raise the temperature of the reaction mixture to 160°C. Then, inject 0.2 mL of Zn(stearate)$_2$ (0.122 g dissolved in 2 mL ODE at 60°C–80°C) and 0.2 mL of sulfur (7 mg dissolved in 2 mL ODE at 60°C–80°C) alternately with a 1-min gap. Repeat this alternate injection of Zn(stearate)$_2$ and sulfur 10 times with a gradual increase of reaction temperature to 200°C. Then, raise the temperature to 220°C and maintain the temperature for 10 min. Then, cool the reaction mixture to room temperature. Typical colloidal solutions of ZnS-capped CdSe with a size-dependent emission property are shown in Figure 3.5.

3.7 HYDROPHOBIC Mn-DOPED ZnS AND Mn-DOPED ZnSeS NANOPARTICLES (4–5 nm)

Take 63 mg Zn(stearate)$_2$, 2 mg Mn(stearate)$_2$, 12 mg S powder and 0.5 g ODA in a three-necked flask and mix with 8 mL octadecene solvent.[16] Purge the reaction mixture with nitrogen/argon for 20 min, then heat to 270°C for 5 min. Next, lower the reaction temperature to 250°C. In a separate vial, dissolve 0.63 g Zn(stearate)$_2$ and 0.25 g stearic acid in 4 mL octadecene by heating at 60°C–80°C, and add the whole solution. Next, heat the mixture for 30 min at 250°C, and finally cool to room temperature. This method produces Mn-doped ZnS of 4–5 nm size with yellow emission under ultraviolet (UV) excitation (Figure 3.6). Synthesis of Mn-doped ZnSeS is similar, except that 150 mg Se powder is used in addition to S. The resultant Mn-doped ZnSeS has yellow emission under UV or blue excitation. Purify the nanoparticles using the standard ethanol-based precipitation and chloroform-based redispersion method (see the description in Section 3.8). Finally, dissolve the particles

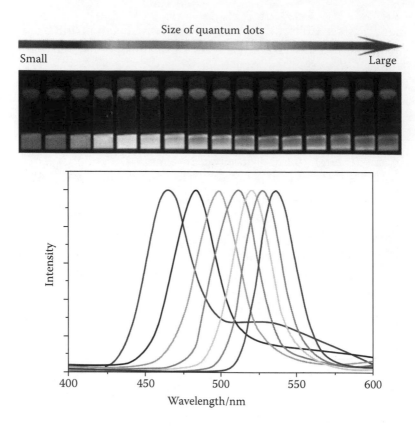

FIGURE 3.5 Colloidal solutions of ZnS-capped CdSe with size-dependent emission property. (Reproduced from Bera, D. et al. 2010. *Materials*, 3, 2260–2345. Copyright by the MDPI, Switzerland.)

in cyclohehane or chloroform with an approximate particle concentration of 0.2–0.5 mg mL^{-1}.

3.8 FLUORESCENT CARBON NANOPARTICLES

3.8.1 Hydrophilic Carbon Nanoparticles (10–50 nm) with Blue-Green Emission

Dissolve 100 mg folic acid in 1 mL oleylamine.[17,18] In another round-bottomed flask, heat 10 mL oleic acid at 280°C in open atmosphere under stirring conditions and inject the folic acid

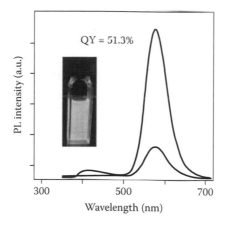

FIGURE 3.6   Emission spectra of colloidal Mn-doped ZnS. (Reprinted with permission Srivastava, B. B. et al. 2010. Highly luminescent Mn-doped ZnS nanocrystals: Gram scale synthesis. *The Journal of Physical Chemistry Letters*, 1, 1454–1458. Copyright 2010 American Chemical Society.)

solution. The solution color changes from yellow to deep brown. Maintain the temperature for 3–5 min and rapidly cool to room temperature. Precipitate the particles by adding acetone and wash using ethanol. Finally, dissolve the particles in water and dialyze using a dialysis membrane (MWCO~12–14 kDa) to remove unreacted reagents/solvents.

3.8.2 Hydrophobic Carbon Nanoparticles with Yellow-Red Emission (3–10 nm)

Dissolve 1 g of ascorbic acid in 12 mL oleylamine and place it in a three-necked round-bottomed flask.[19] Heat the solution at 280°C for 4 h under continuous air flow through the solution using a syringe. Next, stop the reaction and cool to room temperature. Next, purify the hydrophobic nanoparticles using acetone-based precipitation and chloroform-based redispersion. Finally, prepare the nanoparticle solution in chloroform with a concentration of 10–20 mg/mL. Figure 3.7 shows the variable emission property of fluorescent carbon nanoparticles.

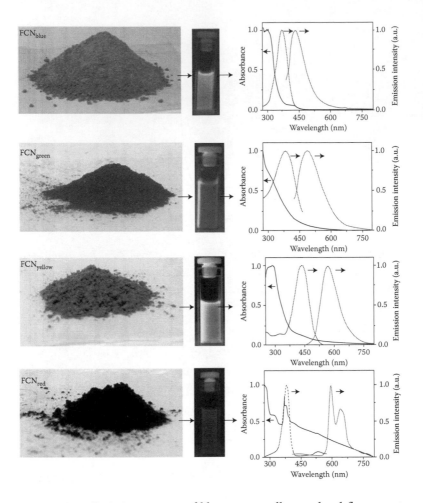

FIGURE 3.7 Emission spectra of blue, green, yellow and red fluorescent carbon nanoparticles. (Reprinted by permission from Springer Nature. *ACS Applied Materials & Interfaces*, 8, 9305–9313. Ali, H. et al., copyright 2016.)

3.9 HYDROPHOBIC IRON OXIDE NANOPARTICLES (5–25 nm)

Dissolve 0.37 g octadecylamine and 0.16 g methyl morpholine N-oxide in 10 mL of octadecene, place the solution in a 250-mL three-necked round-bottomed flask and keep under the magnetic stirring condition.[20,21] Purge the solution with argon/nitrogen for 15–30 min, then heat the solution at 300°C. In a separate vial, dissolve 0.32 g Fe(II)-stearate in 4 mL of hot (50°C–60°C) octadecene and inject the whole solution into the round-bottomed flask. Stop heating after 1–5 min and cool to room temperature. This approach will produce hydrophobic γ-Fe$_2$O$_3$ nanoparticles of 5–6 nm size. If methyl morpholine N-oxide is not used, then Fe$_3$O$_4$ will form with a similar size.

If octadecylamine is replaced with the same equivalent of stearic acid and no methylmorpholine N-oxide is used, then Fe$_3$O$_4$ of 15 nm will be formed in 2 h heating.

If Fe(II)-stearate is replaced with Fe(II)-oleate, octadecylamine is replaced with one equivalent oleic acid and no methylmorpholine N-oxide is used, then Fe$_3$O$_4$ of 10–15 nm will be formed in 30–60 min heating.

If Fe(II)-stearate is replaced with Fe(II)-oleate, octadecylamine is replaced with 2 mL oleic acid and no methylmorpholine N-oxide is used, then Fe$_3$O$_4$ of 25 nm will be formed in 3 h heating. Figure 3.8 shown a TEM image of the iron oxide nanoparticles of varied size that are prepared by this approach.

Purify the nanoparticles as described below before any application. Typically, mix 0.5 mL of nanoparticle solution with 0.5 mL of toluene and 1 mL of acetone. Centrifuge the mixture at 14,000 rpm for 2 min and collect the precipitated particles. Dissolve the precipitate in 1 mL of toluene and precipitate the particles again by adding 1 mL of ethanol. Heat the mixture at 50°C–60°C for 1 min and centrifuge at 25,000 rpm for 2 min. Collect the precipitate, dissolve in 0.5–1.0 mL toluene, precipitate again by adding ethanol and collect the particle by centrifuge. Repeat this precipitation–redispersion procedure two more times, and finally dissolve the particles in 2 mL toluene/cyclohexane prior to different coating.

3.10 HYDROPHOBIC ZnO NANOPARTICLES (5 nm)

Dissolve 220 mg zinc acetate in 20 mL ethanol and add 70 μL oleic acid. In a separate flask, dissolve 360 mg solid tetramethyl

FIGURE 3.8 TEM image of iron oxide nanoparticles of varied size where (a)–(c) represent particle of different size that are prepared in different growth condition. (Reprinted with permission Jana, N. R. et al. 2004. Size- and shape-controlled magnetic (Cr, Mn, Fe, Co, Ni) oxide nanocrystals via a simple and general approach. *Chemistry of Materials*, 16, 3931–3935. Copyright 2004 American Chemical Society.)

ammonium hydroxide in 5 mL ethanol. Heat both solutions to boiling and mix together quickly.[22] Continue heating for exactly 2 min and stop the reaction by adding 50 mL of cold (0°C–10°C) ethanol. A precipitate of ZnO will be observed. Collect the precipitate by centrifuge and redissolve in 10 mL toluene.

3.11 HYDROPHOBIC HYDROXYAPATITE NANORODS/ NANOWIRES (2–5 nm × 10–1000 nm)

Dissolve 120 mg calcium oleate in 8 mL oleic acid and load into a three-necked round-bottomed flask.[23] Purge the solution with argon/nitrogen for 15–20 min under 70°C–80°C. Next, raise the

TABLE 3.1   Reaction Condition-Dependent Length-to-Width
Ratio of Hydroxyapatite Nanorods/Nanowires

Temperature	Reaction Time	Nanorod/Nanowire Size
260°C	30 min	10–12 nm × 2–3 nm
260°C	2 h	20–22 nm × 3–4 nm
300°C	4 h	33–35 nm × 4–5 nm
330°C	2 h	100–1000 nm × 3–4 nm

temperature to 200°C–330°C and inject tetrabutylammonium
phosphate (34 mg dissolved in 1 mL of oleic acid) under stirring
conditions. Maintain the temperature for 15 min to 6 h and
then stop the heating. Nanorods/nanowires will be formed
depending on the reaction temperature and time (see Table 3.1 and
Figure 3.9). Precipitate the nanorods/nanowires by adding acetone
and redisperse the isolated precipitate in chloroform. Perform
another round of purification by ethanol-based precipitation
and chloroform/toluene/cyclohexane based redispersion. Finally,
make the stock solution in chloroform/toluene/cyclohexane.

3.12 HYDROPHOBIC TiO$_2$ NANORODS (2–3 nm × 25–35 nm)

Take 6.0 mL oleic acid in a three-necked round-bottomed flask and
purge with nitrogen for 15–20 min under 70°C–80°C.[24,25] Next,
inject 0.2 mL of titanium isopropoxide under stirring conditions,
and increase the temperature to 300°C. (CAUTION! Titanium
isopropoxide creates fume in open atmosphere, so use a nitrogen/
argon atmosphere while opening the bottle and transferring to the
reaction flask.) Then, inject 0.7 mL of oleylamine under stirring
conditions. Maintain the 300°C temperature for 2 h, then cool to
room temperature. This method produces TiO$_2$ nanorods with
25–35 nm length and 2–3 nm diameter. Precipitate the nanorods
by adding acetone, isolate nanorods by centrifuge and dissolve
them in chloroform. Next, perform a second round of purification
of nanorods by ethanol-based precipitation and chloroform/

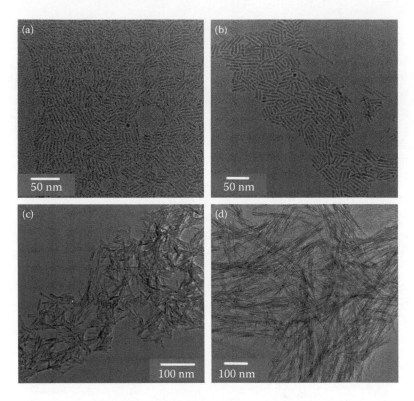

FIGURE 3.9 TEM image of hydroxyapatite nanorods and nanowires of controlled length where (a)–(d) represent nanorod/nanowire of different length, prepared under different growth condition. (Reprinted with permission Das, P. and Jana, N. R. 2016. Length-controlled synthesis of calcium phosphate nanorod and nanowire and application in intracellular protein delivery. *ACS Applied Materials & Interfaces*, 8, 8710–8720. Copyright 2016 American Chemical Society.)

cyclohexane/toluene-based redispersion. Finally, prepare a stock solution of nanorods in 5–6 mL chloroform/cyclohexane/toluene.

3.13 N, F-CODOPED HYDROPHOBIC TiO$_2$ NANOPARTICLES (150–225 nm)

Take 6.0 mL oleic acid in a three-necked round-bottomed flask and purge with nitrogen for 15–20 min under 70°C–80°C.[25] Next, inject

0.2 mL of titanium isopropoxide under stirring conditions and increase the temperature to 300°C. Then, inject 0.7 mL of oleylamine under stirring conditions and maintain the 300°C temperature for 2 h. Next, cool to room temperature and add 360 mg ammonium fluoride and 360 mg urea under magnetic stirring conditions. Raise the temperature to 300°C again under open air, maintain the temperature for 1 h, then stop the heating. As the temperature increases, bubbles will be observed from the solution that decrease within 15–30 min. The solution color changes gradually from white to green within 15–30 min. This method produces N, F-codoped TiO_2 nanoparticles of 150–225 nm size (Figure 3.10). Cool the reaction mixture to room temperature and precipitate nanoparticles by adding acetone. Isolate the particles by centrifuge and dissolve them in chloroform. Next, perform a second round of purification by ethanol-based precipitation and chloroform/cyclohexane/toluene-based redispersion. Finally, prepare a stock solution of particles in 5–6 mL chloroform/cyclohexane/toluene.

3.14 COLLOIDAL GRAPHENE OXIDE VIA MODIFIED HUMMER'S METHOD

Take 200 mg graphite powder and 100 mg sodium nitrate in a beaker and mix with 5 mL concentrated sulfuric acid.[26,27] Cool the mixture to 0°C and add 600 mg of $KMnO_4$ in a stepwise manner under continuous stirring conditions. After half an hour, add 40 mL of water in two steps and cool the mixture to room temperature. Then, add 500 μL of 3% H_2O_2 to consume the excess permanganate. Next, wash the solid with hot water, dry in air and disperse in 20–50 mL distilled water via sonication. Centrifuge the solution at 3000 rpm to remove larger particles and use the supernatant as a colloidal graphene oxide solution with the concentration of 1 mg mL^{-1}.

3.15 POLYASPARTIC ACID MICELLES (25–100 nm)

Suspend 3 g L-aspartic acid in 10 mL of mesitylene under inert conditions, mix with 165 μL of phosphoric acid (88%) and

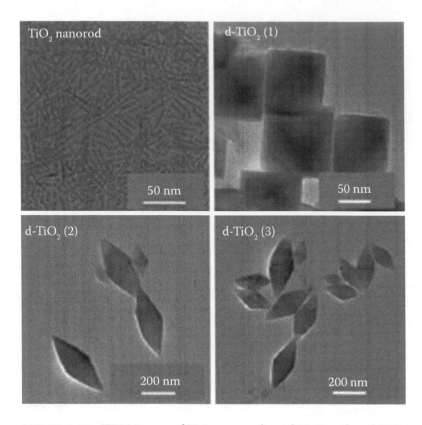

FIGURE 3.10   TEM image of TiO_2 nanorods and N, F-codoped TiO_2 nanoparticles. (Reprinted with permission Biswas, A. et al. 2018. Nitrogen and fluorine codoped, colloidal TiO_2 nanoparticle: Tunable doping, large red-shifted band edge, visible light induced photocatalysis, and cell death. *ACS Applied Materials & Interfaces*, 10, 1976–1986. Copyright 2018 American Chemical Society.)

heat to 150°C for 4 h.[28] Next, collect the white residue at room temperature, dissolve in dimethylformamide and then add excess water to precipitate the polysuccinimide. Wash the precipitate with water to remove phosphoric acid and dimethylformamide and then wash with methanol several times. Finally, dry the solid polysuccinimide in vacuum.

Next, dissolve 250 mg of polysuccinimide in 15 mL of dry dimethylformamide, mix with 135 mg of octadecylamine and heat at 70°C for 24 h under an inert atmosphere. Cool the solution to room temperature, mix with 600 μL PEG-diamine and heat further at 80°C under inert atmosphere for 24 h. Precipitate the resultant material by adding diethyl ether. Wash the precipitate with a methanol–diethyl ether mixture (1:1) several times and dry under vacuum. The solid polymer will form primary amine-terminated micelles once it is dispersed in water, typically in the concentration range of 1–10 mg/mL concentration.

3.16 POLY(LACTIC-CO-GLYCOLIC ACID) NANOPARTICLES AND DRUG-LOADED NANOPARTICLES (100–500 nm)

Poly(lactic-co-glycolic acid) (PLGA) can be purchased from commercial sources. Dissolve 100 mg PLGA in 4 mL methylene chloride.[29,30] In a separate vial, take 100 mL aqueous Tween 80 (1 wt%) solution and add the whole PLGA solution under vigorous stirring conditions for 4 h. In addition, ultrasound can be used to sonicate the mixture. During this time, all the methylene chloride is evaporated. Next, filter the solution through a 0.8-μm membrane filter. Next, centrifuge at 15,000 rpm for 10 min to precipitate the particles. Redisperse the particles in fresh water and use them for further study.

For the preparation of hydrophobic drug-loaded PLGA nanoparticles, separately dissolve 100 mg PLGA in 4 mL methylene chloride and 1–10 mg hydrophobic drug in 0.0–1.0 mL methylene chloride. Mix the two solutions and add to 100 mL aqueous Tween 80 (1 wt%) solution under vigorous stirring conditions. Continue stirring for 4 h; ultrasound can also be used to sonicate the mixture. Next, filter the solution through a 0.8-μm membrane filter, centrifuge at the speed of 15,000 rpm for 10 min to precipitate the particles, redisperse the particles in fresh water and use them for further study.

3.17 LIPOSOME VESICLES (10–25 nm)

Commonly used vesicle components include 1,2-dioleoyl-sn-glycero-3-[phospho-rac-(1- glycerol)] (sodium salt) (DOPG), 1,2-dioleoyl-3-trimethylammonium propane (chloride salt) (DOTAP), 1,2-dipalmitoyl-sn-glycero-3-[phospho-rac-(1-glycerol)] (sodium salt) (DPPG), 1,2-dioleoyl-sn-glycero-3-phosphocholine (DOPC), 1,2-dipalmitoylsn- glycero-3-phosphocholine (DPPC), lyso-palmitoylphosphatidylcholine (lyso-PPC) and cholesterol.[31,32] Prepare a solution of lipid mixture (10–50 mg) in 1–2 mL chloroform. Next, remove the chloroform under a gentle stream of nitrogen. Next, add 2–4 mL of buffer solution and expose under ultrasonication for 30–60 min. Filter through a 0.20 micron and store at 25°C for at least 12 h. This approach can typically produce liposomes with a mean radius of about 10–25 nm.

3.18 ALBUMIN NANOPARTICLES (100 nm)

Prepare 5–10 mL bovine serum albumin (BSA) solution in distilled water with concentration of 1 wt%.[33] Keep the solution under magnetically stirring conditions. Next, adjust the solution pH to 9.0 by adding NaOH and stir the solution overnight. Next, add acetone (desolvating agent) dropwise at a rate of 1 mL/min until the solution becomes just turbid. Finally, add 0.01 mL of a 4% glutaraldehyde–ethanol solution to induce intraparticle cross-linking. Stir the solution continuously at room temperature for 3 h. Purify the BSA nanoparticles by two cycles of centrifugation (12,000–20,000 rpm, 30 min) and redispersion (in 5–10 mL ethanol and ultrasound) to remove unreacted chemicals and free BSA molecules. Store the purified BSA nanoparticles in absolute ethanol at 4°C.

REFERENCES

1. Jana, N. R., Gearheart, L. and Murphy, C. J. 2001. Wet chemical synthesis of high aspect ratio cylindrical gold nanorods. *The Journal of Physical Chemistry B*, 105, 4065–4067.

2. Jana, N. R., Gearheart, L. and Murphy, C. J. 2001. Evidence for seed-mediated nucleation in the chemical reduction of gold salts to gold nanoparticles. *Chemistry of Materials*, 13, 2313–2322.
3. Jana, N. R., Gearheart, L. and Murphy, C. J. 2001. Seeding growth for size control of 5–40 nm diameter gold nanoparticles. *Langmuir*, 17, 6782–6786.
4. Jana, N. R. and Peng, X. 2003. Single-phase and gram-scale routes toward nearly monodisperse Au and other noble metal nanocrystals. *Journal of the American Chemical Society*, 125, 14280–14281.
5. Prasad, B. L. V., Stoeva, S. I., Sorensen, C. M. and Klabunde, K. J. 2002. Digestive ripening of thiolated gold nanoparticles: The effect of alkyl chain length. *Langmuir*, 18, 7515–7520.
6. Zhang, Q., Xie, J., Yang, J. and Lee, J. Y. 2009. Monodisperse icosahedral Ag, Au, and Pd nanoparticles: Size control strategy and superlattice formation. *ACS Nano*, 3, 139–148.
7. Jayesh R. S., Deepti S. S. and Bhagavatula, L. V. P. 2017. Digestive ripening: A fine chemical machining process on the nanoscale. *Langmuir*, 33, 9491–9507.
8. Jana, N. R. 2005. Gram-scale synthesis of soluble, near-monodisperse gold nanorods and other anisotropic nanoparticles. *Small*, 1, 875–882.
9. Sau, T. K. and Murphy, C. J. 2004. Seeded high yield synthesis of short Au nanorods in aqueous solution. *Langmuir*, 20, 6414–6420.
10. Nikoobakht, B. and El-Sayed, M. A. 2003. Preparation and growth mechanism of gold nanorods (NRs) using seed-mediated growth method. *Chemistry of Materials*, 15, 1957–1962.
11. Chang, H.-H. and Murphy, C. J. 2018. Mini gold nanorods with tunable plasmonic peaks beyond 1000 nm. *Chemistry of Materials*, 30, 1427–1435.
12. Macdonald, I. D. G. and Smith, W. E. 1996. Orientation of cytochrome C adsorbed on a citrate-reduced silver colloid surface. *Langmuir*, 12, 706–713.
13. Jana, N. R. 2003. Silver coated gold nanoparticles as new surface enhanced Raman substrate at low analyte concentration. *Analyst*, 128, 954–956.
14. Li, J. J., Wang, Y. A., Guo, W. Z., Keay, J. C., Mishima, T. D., Johnson, M. B. and Peng, X. G. 2003. Large-scale synthesis of nearly monodisperse CdSe/CdS core/shell nanocrystals using air-stable reagents via successive ion layer adsorption and reaction. *Journal of the American Chemical Society*, 125, 12567–12575.

15. Bera, D., Qian, L., Tseng, T.-K., Paul, H. and Holloway, P. H. 2010. Quantum dots and their multimodal applications: A review. *Materials*, 3, 2260–2345.

16. Srivastava, B. B., Jana, S., Karan, N. S., Paria, S., Jana, N. R., Sarma, D. D. and Pradhan, N. 2010. Highly luminescent Mn-doped ZnS nanocrystals: Gram scale synthesis. *The Journal of Physical Chemistry Letters*, 1, 1454–1458.

17. Ali, H., Ghosh, S. and Jana, N. R. 2018. Biomolecule-derived fluorescent carbon nanoparticle as bioimaging probe. *MRS Advances*, 3, 779–788.

18. Bhunia, S. K., Saha, A., Maity, A. R., Ray, S. C. and Jana, N. R. 2013. Carbon nanoparticle-based fluorescent bioimaging probes. *Scientific Reports*, 3, 1473.

19. Ali, H., Bhunia, S. K., Dalal, C. and Jana, N. R. 2016. Red fluorescent carbon nanoparticle-based cell imaging probe. *ACS Applied Materials & Interfaces*, 8, 9305–9313.

20. Jana, N. R., Chen, Y. and Peng, X. 2004. Size- and shape-controlled magnetic (Cr, Mn, Fe, Co, Ni) oxide nanocrystals via a simple and general approach. *Chemistry of Materials*, 16, 3931–3935.

21. Park, J., An, K., Hwang, Y., Park, J. G., Noh, H. J., Kim, J. Y., Park, J.-H., Hwang, N. M. and Hyeon, T. 2004. Ultra-large-scale syntheses of monodisperse nanocrystals. *Nature Materials*, 3, 891–895.

22. Jana, N. R., Yu, H., Ali, E. M., Zheng, Y. and Ying, J. Y. 2007. Controlled photostability of luminescent nanocrystalline ZnO solution for selective detection of aldehydes. *Chemical Communications*, 14, 1406–1408.

23. Das, P. and Jana, N. R. 2016. Length-controlled synthesis of calcium phosphate nanorod and nanowire and application in intracellular protein delivery. *ACS Applied Materials & Interfaces*, 8, 8710–8720.

24. Zhang, Z., Zhong, X., Liu, S., Li, D. and Han, M. 2005. Aminolysis route to monodisperse titania nanorods with tunable aspect ratio. *Angewandte Chemie International Edition*, 44, 3466–3470.

25. Biswas, A., Chakraborty, A. and Jana, N. R. 2018. Nitrogen and fluorine codoped, colloidal TiO_2 nanoparticle: Tunable doping, large red-shifted band edge, visible light induced photocatalysis, and cell death. *ACS Applied Materials & Interfaces*, 10, 1976–1986.

26. Saha, A., Basiruddin, S. K., Ray, S. C., Roy, S. S. and Jana, N. R. 2010. Functionalized graphene and graphene oxide solution via polyacrylate coating. *Nanoscale*, 2, 2777–2782.

27. Mondal, A. and Jana, N. R. 2014. Surfactant-free, stable noble metal–graphene nanocomposite as high performance electrocatalyst. *ACS Catalysis*, 4, 593–599.
28. Chakraborty, A. and Jana, N. R. 2017. Vitamin C-conjugated nanoparticle protects cells from oxidative stress at low doses but induces oxidative stress and cell death at high doses. *ACS Applied Materials & Interfaces*, 9, 41807–41817.
29. Mohammad, A. K. and Reineke, J. J. 2013. Quantitative detection of PLGA nanoparticle degradation in tissues following intravenous administration. *Molecular Pharmaceutics*, 10, 2183–2189.
30. Qin, J., Wei, X., Chen, H., Lv, F., Nan, W., Wang, Y., Zhang, Q. and Chen, H. 2018. mPEG-g-CS-modified PLGA nanoparticle carrier for the codelivery of paclitaxel and epirubicin for breast cancer synergistic therapy. *ACS Biomaterials Science and Engineering*, 4, 1651–1660.
31. Johnsson, M., Silvander, M., Karlsson, G. and Edwards, K. 1999. Effect of PEO-PPO-PEO triblock copolymers on structure and stability of phosphatidylcholine liposomes. *Langmuir*, 15, 6314–6325.
32. Genc, R., Ortiz, M. and O'Sullivan, C. K. 2009. Curvature-tuned preparation of nanoliposomes. *Langmuir*, 25, 12604–12613.
33. Jun, J. Y., Nguyen, H. H., Paik, S. Y., Rju, Chun, H. S., Kang, D.-C. and Ko, S. 2011. Preparation of size-controlled bovine serum albumin (BSA) nanoparticles by a modified desolvation method. *Food Chemistry*, 127, 1892–1898.

CHAPTER 4

Selected Coating Chemistry for Water-Soluble, Core-Shell–Type Nanoparticles with Cross-Linked Shells

4.1 INTRODUCTION

As synthesized nanoparticles need to be protected from aggregation (or growth) and linked with various chemicals/biochemicals, and they should be water dispersible for the intended biomedical application. Coating with molecules or polymer around nanoparticles is the most powerful approach for their protection and functionalization. This chapter summarizes various coating procedures for different nanoparticles. Coating methods are selected based on their general utility to different nanoparticles and advantages in deriving various functional nanoparticles. Attempts have been made to focus on the coating methods that are most reliable and have good reproducibility. Coating methods include silica coating, polyacrylate coating, lipophilic polymer coating, ligand exchange with thiolated molecule/polymer, polyhistidine coating, polydopamine coating and hyperbranched polyglycerol grafting. It is shown that these coating methods produce a variety of primary amine-terminated nanoparticles and nanoparticles with controlled charge/ functional groups. The described coating approaches offer a general guideline to prepare various coated nanoparticles and can be adapted to new nanoparticles of interest.

4.2 SILICA COATING

In this approach, nanoparticles are coated with a silica shell, typically between 1 and 100 nm thickness.[1-3] Typically, methoxy/ ethoxy silanes are used as silica-forming precursors and they are hydrolyzed at the nanoparticle surface to make the silica shell (Figure 4.1). The silica shell thickness can be controlled by using tri-methoxy/ethoxy or tetra-methoxy/ethoxy silane, varying the ratio of nanoparticle to silane, controlling the silane hydrolysis, separating nanoparticles at different coating times and varying the coating time. The surface of the coated nanoparticles can be terminated with primary amine/polyethylene glycol/cation/anion by using different types of silanes. A wide variety of nanoparticles are coated with silica and terminated with primary amine. These coated nanoparticles are described below, as they are very useful for covalent conjugation with various biochemicals.

4.2.1 Hydrophilic Au Nanoparticles (4–5 nm) with Primary Amine Termination

Prepare 10 mL toluene solution of AuCl$_3$ (0.01 M) in the presence of an equimolar amount of didodecyl dimethyl ammonium bromide.[1] Next, add 200 μL toluene solution of (3-mercaptopropyl)

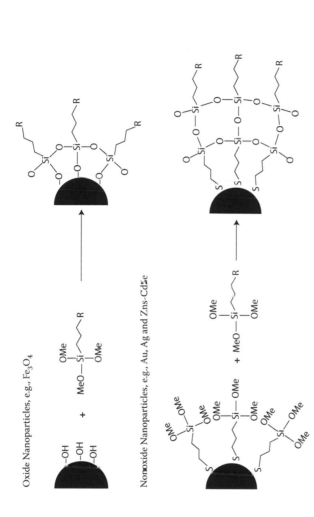

FIGURE 4.1 Approach for making silica-coated metal and metal oxide nanoparticles. (Reprinted with permission from Jana, N. R. et al. 2007. Synthesis of watersoluble and functionalized nanoparticles by silica coating. *Chemistry of Materials*, 19, 5074–5082. Copyright 2007 American Chemical Society.)

trimethoxysilane (0.1 M). In a separate vial, dissolve 25 mg tetrabutyl ammonium borohydride in the presence of 25 mg didodecyl dimethyl ammonium bromide in 1 mL toluene via sonication and inject the whole solution into the gold salt solution under stirring conditions. Gold nanoparticles are formed with the appearance of a pink solution. Next, add 2 mL of toluene solution of N-(2-aminoethyl)-3-aminopropyltrimethoxysilane (0.1 M), and heat the resulting solution to 50°C–60°C. Particles began to precipitate after 3–4 min and completely precipitate after 15 min. Wash the precipitate twice with toluene, twice with chloroform and twice with ethanol. Next, dissolve the particles in 2 mL water with the formation of a dark-brown solution. In this process, the size of Au nanoparticles would be 4–5 nm and the overall hydrodynamic size would be 6–10 nm.

4.2.2 Hydrophilic Ag Nanoparticles (4–5 nm) with Primary Amine Termination

Dissolve 17 mg silver acetate and 72 mg dodecylamine in 10 mL toluene via sonication.[1] An optically clear solution is formed. Next, add 200 μL toluene solution of (3-mercaprtopropyl) trimethoxysilane (0.1 M). In a separate vial, dissolve 25 mg tetrabutyl ammonium borohydride and 17 mg decanoic acid in 1 mL toluene (CAUTION! *Be careful of hydrogen gas formation during mixing*) and inject the whole solution into the earlier solution. The solution color appears dark yellow due to the formation of colloidal silver nanoparticles. Next, add 2 mL of the toluene solution of N-(2-aminoethyl)-3-aminopropyltrimethoxysilane (0.1 M), and heat the entire solution to 65°C for 15–30 min or until complete precipitation. Next, centrifuge the solution and remove the supernatant. Collect the precipitate; wash repeatedly with toluene, chloroform and ethanol; and finally dissolve the particles in 2 mL distilled water with the formation of a brown colloidal solution. In this process, the core size of the Ag nanoparticles would be 3–5 nm and the overall size would be 9–12 nm.

4.2.3 Hydrophilic Iron Oxide Nanoparticles (5 nm) with Primary Amine Termination

Mix 2 mL toluene solution of purified nanoparticles (see Section 3.9) with 1.0 mL toluene solution of N-(2-aminoethyl)-3-aminopropyltrimethoxy silane (0.01 M) and 1 mL methanol solution of tetramethyl ammonium hydroxide (0.01 M).[1] Stir the mixture using a magnetic stirrer and heat to 80°C for 10–20 min until complete precipitation of particles. Next, cool the solution and discard the supernatant. Wash the solid precipitate repeatedly with toluene and/or ethanol and then dissolve in 1 mL water. If the precipitate is partially soluble, a drop of aqueous formic acid solution can be added for complete solubilization.

4.2.4 Hydroxyapatite Nanorods with Primary Amine Termination

Take 3 mg hydroxyapatite nanorod and dissolve in 2 mL toluene[2] (see Section 3.11). In a separate vial, dissolve 30 μL of N-(2-aminoethyl)-3-aminopropyltrimethoxysilane in 2 mL of toluene and mix with nanorod solution. Keep the solution under stirring conditions and raise the temperature to 70°C–80°C. In another vial, mix 15 μL of tetramethylammonium hydroxide solution in 0.5 mL of methanol and add dropwise into the nanorod solution. Continue heating at 70°C–80°C for 20–30 min until all the nanorods precipitate. Wash the precipitate with toluene and ethanol, and finally dissolve in 1–2 mL distilled water.

4.2.5 Primary Amine-Terminated Magnetic Mesoporous Silica Nanoparticles (50–100 nm)

Take 2 mL colloidal solution of silica-coated iron oxide nanoparticles (see Section 4.2.3) and mix with 45 mL distilled water and 5 mL cetyltrimethylammonium bromide (0.015 M) solution under stirring conditions. After 15 min, add 1.5 mL NH₃ solution (25%).[3] Next, add 0.5 mL ethanolic solution of tetraethylorthosilicate (300 μL dissolved in 2.5 mL ethanol), 2 mL ethanolic solution of N-(2-aminoethyl)-3-aminopropyltrimethoxy silane (50 μL

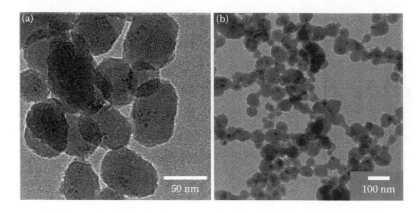

FIGURE 4.2 TEM image of magnetic mesoporous silica nanoparticles where a, b represent images at different magnifications. (Reprinted with permission from Sinha, A. et al. 2015. β-cyclodextrin functionalized magnetic mesoporous silica colloid for cholesterol separation. *ACS Applied Materials & Interfaces*, 7, 1340–1347. Copyright 2015 American Chemical Society.)

dissolved in 10 mL ethanol) and 2 mL dimethylformamide in a stepwise manner at 5-min intervals. Keep the whole solution under magnetic stirring conditions for 3 h. Next, precipitate the particles by adding excesses ethanol. Isolate the particles by centrifuge and wash three times with ethanol and three times with water. Next, disperse the particles in ethanolic solution of NH_4NO_3 (250 mg NH_4NO_3 dissolved in 25 mL ethanol) by sonication and heat the whole solution to 80°C for 2 h under stirring conditions. This process extracts the surfactant from porous particles, and the process may be repeated twice. Finally, disperse the particles in 5–10 mL in water and use as stock solution for further use. A typical transmission electron microscopy (TEM) image of the particles is shown in Figure 4.2.

4.3 POLYACRYLATE COATING

In this approach, nanoparticles are coated with a polymeric shell of polyacrylate, typically of 5–50 nm thickness.[4–9] Acryl monomers are used as polyacrylate precursors and are polymerized at the nanoparticle surface (Figure 4.3). The polyacrylate shell thickness can be controlled by varying the ratio of nanoparticle to acryl monomers and the coating time. Polyacrylate-coated nanoparticles can be terminated with primary amine/polyethylene glycol and their surface charge can be tuned to cationic/anionic/zwitterionic by using different types of functional acrylates. A wide variety of nanoparticles are coated with polyacrylate with very high colloidal stability and terminated with primary amine. These coated nanoparticles are described below, as they are very useful for covalent conjugation with various biochemicals.

4.3.1 Hydrophilic Pegylated Nanoparticles with Cationic Surface Charge and Primary Amine Termination

This method is applicable to all the hydrophobic nanoparticles described in Chapter 3 (except thiol-capped hydrophobic nanoparticles and hydrophobic carbon nanoparticles).[4-9] Prepare 12 mL Igepal-cyclohexane reverse micelle solution by mixing 9 mL toluene with 3 mL Igepal surfactant. In this solution, mix 100 μL aqueous solution of N-(3-aminopropyl) methacrylamide hydrochloride (18 mg dissolved in 100 μL water), 100 μL aqueous solution of poly(ethylene glycol) methacrylate (average M_n 360, 36 μL dissolved in 100 μL water) and 100 μL aqueous solution of methylene-bis-acrylamide (3 mg dissolved in 100 μL water by 10 min sonication). Next, add 2 mL cyclohexane solution of nanoparticles and 100 μL tetramethyl ethylenediamine. Place this solution in a three-necked flask, put under inert atmosphere by purging nitrogen for 20 min and keep under magnetically stirring conditions. Finally, add ammonium persulfate solution (3 mg dissolved in 100 μL water) to initiate polymerization. Continue polymerization for 15–60 min, then precipitate the particles by adding ethanol. Wash the particles with chloroform and ethanol and finally dissolve in 2 mL distilled water. If you take 2–5 nm hydrophobic nanoparticles, use 15 min coating time to prepare particles of 10 nm hydrodynamic size, 30 min coating time to

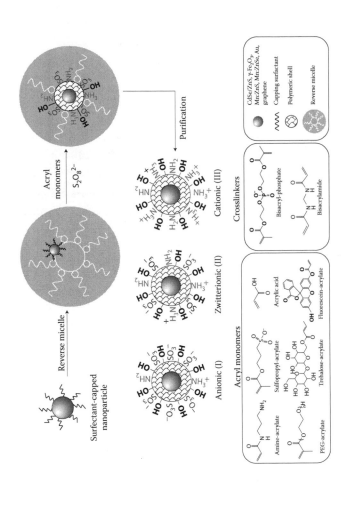

FIGURE 4.3 Schematic representation of polyacrylate coating on hydrophobic nanoparticle surface in reverse micelle. (Reprinted with permission from Chakraborty, A. et al. 2018. Colloidal nanobioconjugate with complementary surface chemistry for cellular and subcellular targeting. *Langmuir*, 34, 13461–13471. Copyright 2018 American Chemical Society.)

prepare particles of 20 nm hydrodynamic size and 60 min coating time for 40 nm hydrodynamic size.

4.3.2 Hydrophilic Nanoparticles with Anionic Surface Charge and Primary Amine Termination

This method is applicable to all the hydrophobic nanoparticles described in Chapter 3 (except thiol-capped hydrophobic nanoparticles and hydrophobic carbon nanoparticles).[4-9] Prepare 12 mL of Igepal-cyclohexane reverse micelle solution by mixing 9 mL cyclohexane with 3 mL Igepal surfactant. To this solution, mix 100 μL aqueous solution of N-(3-aminopropyl) methacrylamide hydrochloride (1.8 mg dissolved in 100 μL water), 100 μL aqueous solution of 3-sulfopropyl methyl acrylate (22 mg dissolved in 100 μL water) and 100 μL aqueous solution of methylene-bis-acrylamide (3 mg dissolved in 100 μL water by 10 min sonication). Next, add 2 mL cyclohexane solution of nanoparticles and 100 μL tetramethyl ethylenediamine. Place this solution in a three-necked flask, put under inert atmosphere by purging nitrogen for 20 min and keep under magnetically stirring conditions. Finally, add ammonium persulfate solution (3 mg dissolved in 100 μL water) to initiate the polymerization. Continue polymerization for 15–60 min, then precipitate the particles by adding ethanol. Wash the particles with chloroform and ethanol and finally dissolve in 2 mL distilled water. If you take 2–5 nm hydrophobic nanoparticles, use 15 min coating time for preparing 10 nm hydrodynamic size, 30 min coating time for 20 nm hydrodynamic size and 60 min coating time for 40 nm hydrodynamic size.

4.3.3 Pegylated Hydrophilic Nanoparticles with Near-Zero Surface Charge and Primary Amine Termination

This method is applicable to all the hydrophobic nanoparticles described in Chapter 3 (except thiol-capped hydrophobic nanoparticles and hydrophobic carbon nanoparticles).[4-9] Prepare 12 mL of Igepal-cyclohexane reverse micelle solution as described above and mix with 100 μL aqueous solution of N-(3-aminopropyl) methacrylamide hydrochloride (2 mg dissolved in 100 μL water), 100 μL aqueous solution of poly(ethylene glycol) methacrylate (average M_n 360, 36 μL dissolved in 100 μL water) and 100 μL aqueous solution of methylene-bis-acrylamide (3 mg dissolved in 100 μL water by 10 min sonication). Next, add 2 mL cyclohexane solution of nanoparticles and 100 μL of tetramethyl ethylenediamine. Place this solution in a three-necked flask, put under inert atmosphere by purging nitrogen for 20 min and keep under magnetically stirring conditions. Finally, add ammonium persulfate solution (3 mg dissolved in 100 μL water) to initiate polymerization. Continue polymerization for 15–60 min, then precipitate the particles by adding ethanol. Wash the particles with chloroform and ethanol and finally dissolve in 2 mL distilled water. If you take 2–5 nm hydrophobic nanoparticles, use 15 min coating time for 10 nm hydrodynamic size, 30 min coating time for 20 nm hydrodynamic size and 60 min coating time for 40 nm hydrodynamic size.

4.3.4 Hydrophilic Nanoparticles with Zwitterionic Surface Charge and Primary Amine Termination

This approach can be used for all the hydrophobic nanoparticles described in Chapter 3 (except thiol-capped hydrophobic nanoparticles and hydrophobic carbon nanoparticles).[4–9] Prepare 10 mL reverse micelle solution by mixing 2.5 mL Igepal with 7.5 mL cyclohexane. Take four separate vials and dissolve 4–10 mg N-(3-aminopropyl)methacrylamide hydrochloride in 0.1 mL water, 12–25 mg 3-sulfopropyl methyl acrylate in 0.1 mL water, 0.036 mL poly(ethylene glycol) methacrylate (average M_n 360) in 0.1 mL water and 0.006 mL (bis[2-(methacryloyloxy)ethyl] phosphate) in 0.1 mL water. Add 1–2 mL reverse micelles to each solution to make an optically clear solution. Take a three-necked flask with 2 mL nanoparticle solution and mix all four acrylate solutions. Purge with nitrogen for 15 min. Next, add 0.1 mL tetramethylethylenediamine and 0.1 mL aqueous solution of 3 mg ammonium persulfate.

Continue the polymerization for 15–60 min under an oxygen-free atmosphere. Next, stop the reaction by adding absolute alcohol that precipitates the particles. Wash the particles repeatedly with chloroform and ethanol, then dissolve in 5 mL of water. If you take 2–5 nm hydrophobic nanoparticles, use 15 min coating time for preparing particles of 10 nm hydrodynamic size, 30 min coating time to prepare particles of 20 nm hydrodynamic size and 60 min coating time for 40 nm hydrodynamic size.

4.3.5 Plasmonic-Fluorescent, Hydrophilic Nanoparticles with Primary Amine Termination

This approach can be used for the hydrophobic Au/Ag nanoparticles described in Sections 3.2.4 and 3.4.2 (except thiol-capped hydrophobic Au/Ag nanoparticles)[10] (Figure 4.4). Prepare 10 mL reverse micelle solution by mixing 2.5 mL Igepal with 7.5 mL cyclohexane. Take four separate vials and dissolve 13 mg N (3-aminopropyl)methacrylamide hydrochloride in 0.1 mL water, 36 mL poly(ethylene glycol) methacrylate (average M_n 360) in 0.1 mL water, 3 mg fluorescein o acrylate in 0.1 mL tetramethyl ethylenedimamine and 3 mg methylene-bis-acrylamine in 0.1 mL water. Add 1–2 mL reverse micelles to each solution to make an optically clear solution. Take a three-necked flask with 2 mL nanoparticle solution and mix all four acrylate solutions. Purge with nitrogen for 15 min, then add 0.1 mL aqueous solution of 3 mg ammonium persulfate. Continue the polymerization for 60 min under oxygen-free atmosphere. Next, stop the reaction by adding absolute alcohol that precipitates the particles. Wash the particles repeatedly with chloroform and ethanol, then dissolve in 2–3 mL of water. Dialyze the solution overnight against basic water using the molecular weight cutoff of membrane of 12,000–14,000.

4.3.6 Magnetic-Fluorescent, Hydrophilic Nanoparticles with Primary Amine Termination

This approach can be used for the hydrophobic iron oxide nanoparticles described in Section 3.9.[11] Prepare 10 mL reverse

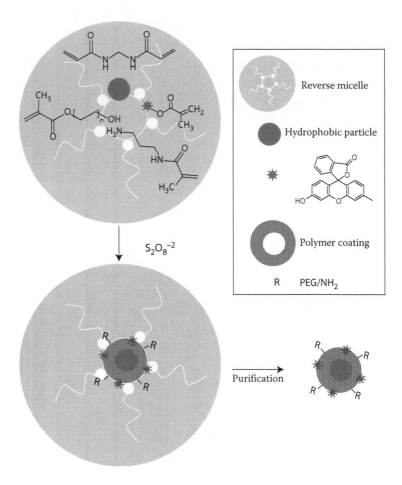

FIGURE 4.4 Schematic representation of making plasmonic-fluorescent nanoparticles via polyacrylate coating. (Reprinted with permission from Saha, A. et al. 2009. Functionalized plasmonic-fluorescent nanoparticles for imaging and detection. *The Journal of Physical Chemistry C*, 113, 18492–18498. Copyright 2009 American Chemical Society.)

micelle solution by mixing 2.5 mL Igepal with 7.5 mL cyclohexane. Take four separate vials and dissolve 13 mg *N*-(3-aminopropyl) methacrylamide hydrochloride in 0.1 mL water, 36 mg poly(ethylene glycol) methacrylate (average M_n 360) in 0.1 mL water, 3 mg fluorescein o-acrylate in 0.1 mL tetramethyl ethylenediamine and

3 mg methylene-bis-acrylamide in 0.1 mL water. Add 1–2 mL reverse micelles to each solution to make an optically clear solution. Take a three-necked flask with 2 mL nanoparticle solution and mix all four acrylate solutions. Purge with nitrogen for 15 min. Next, add 0.1 mL aqueous solution of 3 mg ammonium persulfate. Continue the polymerization for 60 min under an oxygen-free atmosphere. Next, stop the reaction by adding absolute alcohol that precipitate the particles. Wash the particles repeatedly with chloroform and ethanol, then dissolve in 2–3 mL of water. Dialyze the solution overnight to remove all the free reagents.

4.3.7 Hydrophilic Gold Nanorods with Cationic Surface and Primary Amine Termination

Take four microcentrifuge tubes, each containing 1.5 mL of as-synthesized gold nanorod solution (see Section 3.3.1).[12] Centrifuge at 16,000 rpm for 5 min, collect the precipitates from the four microcentrifuge tubes and dissolve in 500 μL CTAB (0.2 M) solution. In a separate vial, prepare 10 mL Igepal cyclohexane reverse micelles by mixing 2.5 mL of Igepal with 7.5 mL of cyclohexane. Next, use three separate vials to prepare reverse micelle solution of acryl monomers. In the first vial, dissolve 24 mg of N-(3-aminopropyl)acryl amide in 0.1 mL water and mix with 2 mL reverse micelles. In the second vial, dissolve 36 μL of poly(ethylene glycol) methacrylate (average M_n 360) in 0.1 mL water and mix with 2 mL reverse micelles. In the third vial, dissolve 3 mg N,N'-methylene-bis-acrylamide in 0.1 mL water and mix with 2 mL reverse micelles. Mix all three acryl monomer solutions and transfer them into a three-necked flask, and add 3–4 mL of reverse micelle solution further. Then, add 0.5 mL of concentrated gold nanorod solution to this mixture, followed by the addition of 100 μL of N,N,N',N'-tetramethyl ethylenediamine. Keep this optically clear solution in magnetic stirring conditions and purge with nitrogen for 15 min to make the reaction mixture O_2-free. After that, add ammonium persulfate solution (3 mg dissolved in 100 μL of H_2O) to initiate the polymerization process. Continue the reaction for 1 h in an N_2 atmosphere, then quench the reaction by adding a small amount of ethanol that precipitates the nanorods. Wash the precipitate with chloroform, and finally dissolve in water.

4.3.8 Colloidal Graphene Oxide and Reduced Graphene Oxide with Primary Amine Termination

Take four microcentrifuge tubes with 2 mL volume. In one of them, dissolve 1.8 mg N-(3-aminopropyl) methacrylamide hydrochloride in 0.1 mL water.[13] In the second vial, dissolve 0.03 mL polyethyleneglycol methacrylate (average M_n 360) in 0.1 mL water. In the third vial, dissolve 0.005 mL bis[2-(methacryloyloxy)-ethyl] phosphate in 0.1 mL of water. In the fourth vial, dissolve 24 mg 3-sulfopropyl methacrylate potassium salt in 0.1 mL water. Next, add 0.5 mL of Igepal and 1.4 mL of cyclohexane to each microcentrifuge tube and shake vigorously to make them optically clear. Next, take a three-necked flask and mix all the solutions. Then, add 500 μL of concentrated aqueous graphene oxide solution and 100 μL of organic base (N,N,N,N-tetramethyl ethylenediamine). The whole solution should be optically clear. Keep the solution under magnetic stirring conditions and purge with nitrogen for 15 min to make the reaction mixture O_2 free. After that, add ammonium persulfate solution (3 mg dissolved in 100 μL of H_2O) to the reaction mixture. Continue the reaction for 1 h in N_2 atmosphere, then add a small amount of ethanol to precipitate the particles. Wash the precipitates repeatedly with chloroform and ethanol and dissolve in doubly distilled water. In order to remove the unbound polymer, centrifuge this solution at 16,000 rpm for 10 min and redisperse the precipitated particles in 2–5 mL fresh water. Use this solution for further applications.

In order to prepare a reduced graphene oxide solution, take 10 mL of coated graphene oxide solution (1 mg/mL) in a glass vial and add 0.1 mL hydrazine hydrate (1% of the original volume). Heat the solution at 70°C for 30 min under stirring conditions. The solution color slowly turns black during this time. Keep this solution at room temperature overnight, then dialyze overnight to remove free hydrazine. Use the dialyzed solution for further applications.

4.4 LIPOPHILIC POLYMER COATING

In this approach, hydrophobic nanoparticles are coated with a monolayer of lipophilic/amphiphilic polymer.[14-20] This approach utilizes the strong interaction between the self-assembled surfactant monolayer around hydrophobic nanoparticles and the lipophilic/amphiphilic component of polymer. The most commonly used polymer is poly(maleic-*alt*-1-octadecene), and when this polymer is mixed with hydrophobic nanoparticles, the octadecene components form an interdigited bilayer with the surfactant monolayer of nanoparticles, leaving its polar components in an outward direction (Figure 4.5). The polymer shell can be crosslinked and functionalized further. This type of polymer coating is applied to a wide variety of hydrophobic nanoparticles. Described below are the lipophilic/amphiphilic polymer coating approaches that provide primary amine termination or functionalization with various biochemicals.

4.4.1 Primary Amine-Terminated Hydrophilic Nanoparticles via Polymaleic Anhydride Coating of Hydrophobic Nanoparticles

This approach can be applied to all the hydrophobic nanoparticles described in Chapter 3. Dissolve 20 mg of poly(maleic-*alt*-1-octadecene) in 1 mL chloroform.[14-18] Next, add 1–2 mL chloroform solution of nanoparticle and sonicate for 10 min. In a separate vial, dissolve 80 μL polyethylene glycol (PEG)–diamine (*O,O'*-bis(2-amino propyl) polypropylene glycol-*block*-polyethylene glycol-*block*-polypropylene glycol) in 1 mL chloroform and add 300 μL of it, followed by sonication for 10 min. Then, add the remaining 700 μL PEG–diamine solution and sonicate for 10 min. Keep the solution overnight for complete evaporation of chloroform. Next, dissolve the precipitate in 2 mL aqueous sodium carbonate solution by 1–2 min sonication. Use a low-speed centrifuge to

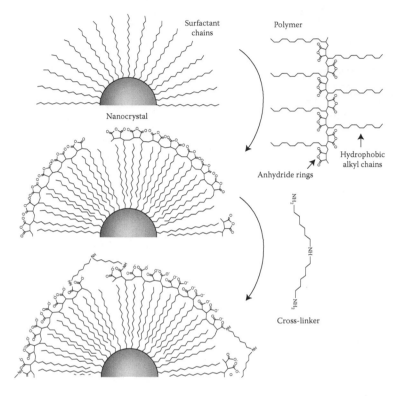

FIGURE 4.5   Schematic representation of polymaleic anhydride coating around hydrophobic nanoparticles. (Reprinted with permission from Pellegrino, T. et al. 2004. Hydrophobic nanocrystals coated with an amphiphilic polymer shell: A general route to water soluble nanocrystals. *Nano Letters*, 4, 703–707. Copyright 2004 American Chemical Society.)

remove larger nanoparticles and collect the supernatant as a stock solution of nanoparticles.

4.4.2 One-Step Polymaleic Anhydride Coating and Functionalization of Hydrophobic Nanoparticles

This coating approach can be used for all the hydrophobic nanoparticles described in Chapter 3[18] (Figure 4.6). Dissolve 20 mg of poly(maleic anhydride alt-1-octadecene) in 1 mL of chloroform through sonication. Next, add 1–2 mL of purified hydrophobic

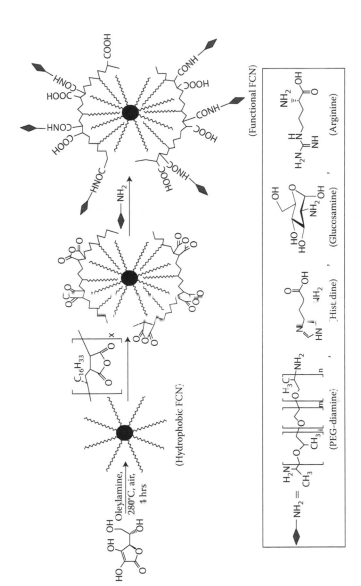

FIGURE 4.6 Schematic representation of one-step poly maleic anhydride coating and functionalization of hydrophobic nanoparticles. (Reprinted with permission from Ali, H. et al. 2016. Red fluorescent carbon nanoparticle-based cell imaging probe. *ACS Applied Materials & Interfaces*, 8, 9305–9313. Copyright 2016 American Chemical Society.)

nanoparticle solution and sonicate the mixture (typically for 5–10 min) until a clear solution appears. Next, evaporate the chloroform with the formation of solid residue. In a separate vial, dissolve the primary amine-containing molecule (e.g., PEG-amine, glucosamone, arginine, histidine) in a carbonate buffer solution of pH 9.5 with a concentration of 5–10 mg/mL. Add 2 mL of this solution to the above-mentioned solid residue and stir the solution mixture for 4–5 h until all particles dissolve. Centrifuge at 15,000 rpm for 20 min to separate larger particles or unreacted polymer. Finally, dialyze the solution using a dialysis tube (molecular weight cutoff: ~12,000 Da) to separate unbound molecules and reagents. The resultant nanoparticle solution is then used as a stock solution.

4.4.3 One-Step Polyaspartic Acid Coating and Functionalization of Hydrophobic Nanoparticles

This approach is also applicable to all the hydrophobic nanoparticles described in Chapter 3[19] (Figure 4.7). Mix 1.0 mL of chloroform solution of nanoparticles with 1.0 mL chloroform solution of lipophilic polyaspartic acid (see the synthesis method described in Section 3.15) and sonicate for 15 min. In a separate vial, take 10 mL hot (50°C–60°C) aqueous bicarbonate (pH ~ 10) solution of biochemical under vigorous stirring conditions. The biochemical may be arginine, spermine, glucosamine or cysteamine. To this solution, add the nanoparticle-polyaspartic acid solution dropwise within 10 min. Wait for complete evaporation of chloroform within 10 min, with the resultant formation of the aqueous nanoparticle solution. Cool the solution to room temperature and dialyze against distilled water using a dialysis membrane (molecular weight cutoff: 12,000–14,000).

4.4.4 Primary Amine-Terminated Colloidal Graphene Oxide and Reduced Graphene Oxide via Polymaleic Anhydride Coating

First, transform hydrophlilic graphene oxide to hydrophobic graphene oxide as described below[20] (Figure 4.8). Take 5 mL

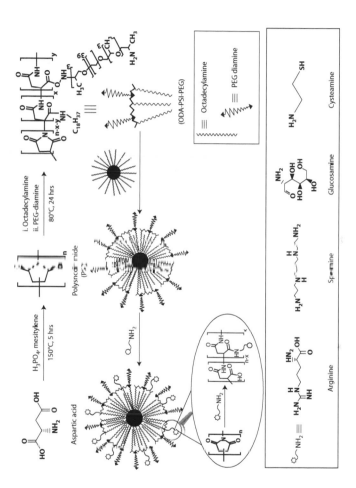

FIGURE 4.7 Schematic representation of one-step poly(aspartic acid coating and functionalization of hydrophobic nanoparticles. (Reprinted with permission from Debnath, K. et al. 2016. Phase transfer and surface functionalization of hydrophobic nanoparticle using amphiphilic poly(amino acid). *Langmuir*, 32, 2798–2807. Copyright 2016 American Chemical Society.)

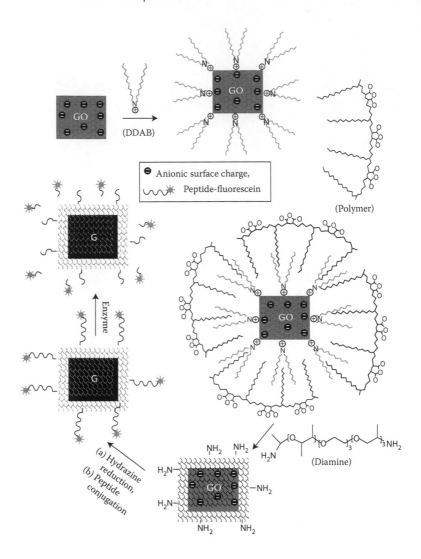

FIGURE 4.8 Schematic representation of making primary amine-terminated colloidal graphene oxide and reduced graphene oxide via polymaleic anhydride coating. (Reprinted with permission from Bhunia, S. K. and Jana, N. R. 2011. Peptide-functionalized colloidal graphene via interdigited bilayer coating and fluorescence turn-on detection of enzyme. *ACS Applied Materials & Interfaces*, 3, 3335–3341. Copyright 2011 American Chemical Society.)

graphene oxide solution (see Section 3.14), add 10 mg didodecyl dimethyl ammonium bromide and sonicate for 7 min. Next, adjust the pH to 8 by adding 75 mg of Na_2HPO_4, followed by 5 min sonication. Next, add 3 mL of $CHCl_3$, stir for a few seconds and keep the solution untouched for phase separation. The colorless $CHCl_3$ phase becomes brown, leaving the colorless aqueous phase, suggesting that graphene oxide is transferred from the aqueous to the organic phase. Collect the bottom organic phase and mix it with $CHCl_3$ solution of polymaleic anhydride-alt-1-octadecene (40 mg dissolved in 0.5 mL $CHCl_3$) and sonicate for 5 min. Next, add $CHCl_3$ solution of PEG-diamine [O,O'-bis(2-aminopropyl)polypropylene glycol-block-polyethylene glycol-block-polypropylene glycol] (60 µL dissolved in 500 µL $CHCl_3$) in two steps. First, add 100 µL and sonicate for 1 min, then add the rest of the solution after 20 min. Evaporate the $CHCl_3$ and mix the brownish precipitate with 3 mL of water and 5 mg of Na_2CO_3. Sonicate the whole mixture for 5–10 min and wait overnight. An optically clear brown solution will form that can be used as a stock solution.

Purify the polymer-coated graphene oxide from the free polymer using acetone-induced precipitation. Typically, mix 1.0 mL of the polymer-coated solution with 1.0 mL acetone and centrifuge at 7000 rpm for 1 min. Collect the brown precipitate, and dissolve in 200 µL of fresh water. Next, add 600 µL acetone and again centrifuge at 10,000 rpm for 2 min. Collect the brown precipitate, dissolve in 2.5 mL of fresh water and use for further experiments.

Convert the polymer-coated graphene oxide to polymer-coated reduced graphene oxide via hydrazine-based reduction. Typically, add 40 µL of hydrazine hydrate to the 2.5 mL solution of polymer-coated graphene oxide and heat to 70°C–80°C for 1 h. A deep black solution will appear. Dialyze the solution overnight to remove excess hydrazine and use it as a stock solution.

4.5 LIGAND EXCHANGE APPROACH

This approach utilizes the strong chemical affinity of thiol to some specific nanoparticles such as Au, Ag and QDs. In particular, nanoparticles that are capped with weakly adsorbed molecules/surfactants are exchanged with thiolated small molecules or polymers.[21-23] Although this is a simple approach, it is limited to selected nanoparticles and conjugation chemistries. It has been shown that multidentate thiols offer more stable coating than monodentate thiols; thus, thiolated polymers are used for coating and functionalization of nanoparticles. Some of them are described below.

4.5.1 Ligand Exchange of Hydrophobic Nanoparticles by Thiolated Polyaspartic Acid or Thiolated Chitosan Oligosaccharide

This approach can be applied to hydrophobic Au/Ag nanoparticles and quantum dots.[21,22] Prepare a reverse micelle by mixing 0.5 mL Igepal with 1.5 mL cyclohehane. Next, add 0.5–1.0 mL hydrophobic Au/Ag/quantum dot solution. In a separate vial, dissolve 10 mg thiolated polymer in 0.1 mL water and mix with the nanoparticle solution via sonication. After 5–10 min, add a few drops of ethanol to precipitate the nanoparticles. Wash with ethanol, and finally dissolve the particles in 1–2 mL water.

4.5.2 Ligand Exchange of Hydrophilic Nanoparticles by Thiolated Polyaspartic Acid or Thiolated Chitosan Oligosaccharide

This approach can be used for water-soluble Ag/Ag nanoparticles capped with citrate or cetyltrimethylamonium bromide surfactant.[21,22] First, prepare a nanoparticle dispersion free from excess surfactant. Dissolve 5–10 mg polymer in 0.5–1.0 mL water. Next, add 1–2 mL aqueous solution of nanoparticles and sonicate for 5–10 min. Next, separate the particles from the free polymers by centrifuge and redisperse the particles in fresh water.

4.5.3 Ligand Exchange of Hydrophobic Nanoparticles by Thiolated Small Molecules

This approach can be used for hydrophobic Ag/Ag nanoparticles capped with fatty acid/amines described above (see Sections 3.2.4

and 3.4.2).[23] Most common thiols include mercaptoundecanoic acid, mercaptopropanoic acid and thiolated PEG. First, prepare an ethanolic solution of thiol (0.1 M). Next, mix 0.1–0.2 mL of this solution with 1–2 mL toluene solution of nanoparticles and shake for 5–10 min. Usually, particles start to precipitate from the solution. Otherwise, add ethanol to induce particle precipitation. Next, separate the particles by centrifuge and redisperse the particles in fresh water.

4.6 POLYHISTIDINE COATING FOR MAKING PRIMARY AMINE-TERMINATED NANOPARTICLES

This approach can be used for all the hydrophobic nanoparticles described in Chapter 3 (except thiol-capped nanoparticles and hydrophobic carbon nanoparticles).[24,25] In this approach, nanoparticles are coated with imidazole-based polymer via the reaction of imidazole nitrogen groups with alkyl bromide.[24,25] Two types of bromides, 1,4-dibromo-2,3-butanediol and 2,4,6 tris(bromomethyl)mesitylene, are commonly used at an optimum ratio, where dibromobutanol introduces water-soluble alcohol groups and tribromomesitylene cross-link the polymer. The nature of imidazole dictates the particle surface charge and chemical functionality. In particular, the use of histidine as an imidazole precursor offers a biocompartible polymer coating with primary amine termination.[25] The polymer coating is performed in an Igepal–cyclohexane reverse micelle medium where hydrophobic nanoparticles and polymer-forming precursors can be solubilized. A typical procedure is described below.

Prepare 12 mL of Igepal-cyclohexane reverse micelle by mixing 3 mL of Igepal with 9 mL cyclohexane. Dissolve 10 mg of 2,4,6-tris(bromomethyl)mesitylene in the above-mentioned Igepal-cyclohexane reverse micelle and mix with 1–2 mL nanoparticle solution. In another vial, dissolve 8 mg histidine in 100 μL of water and mix with the earlier solution. Finally, add 6 μL of N,N,N,N-tetramethylethylenediamine base to the reaction mixture, and continue the reaction overnight under stirring. Next, precipitate the particles by adding ethanol, repeatedly wash with ethanol and finally dissolve in 5–10 mL water. The aqueous solution may be further purified by dialysis against distilled water using a dialysis membrane (molecular weight cutoff: 12,000–14,000 Da)

4.7 HYPERBRANCHED POLYGLYCEROL GRAFTING OF PRIMARY AMINE-TERMINATED NANOPARTICLES

Hyperbranched polyglycerol grafting is particularly useful to increase the colloidal stability of silica-coated nanoparticles.[26] However, the approach can be adapted to any primary amine-terminated nanoparticle. The method involves the reaction of primary amines at the nanoparticle surface with glycidol, initiating ring-opening polymerization and generation of hyperbranched polyglycerol[26] (Figure 4.9). A typical procedure is described below. Disperse 10–50 mg silica-coated nanoparticles in 10–50 mL dry dimethylformamide, mix with 70 mg imidazole and heat the solution at 90°C–100°C under an argon atmosphere. Next, gradually add 2 mL dimethylformamide solution of glycidol (500–750 μL of glycidol dissolved in 2 mL dry dimethylformamide) within 15–20 min and heat the mixture for 6–7 h. Cool the solution and precipitate the particles by adding 5–20 mL of diethyl ether. Isolate the particles via centrifuge, repeatedly wash the particles with diethyl ether and then dissolve in 5–10 mL distilled water.

4.8 POLYDOPAMINE COATING

This coating approach involves oxidative self-polymerization of dopamine under mild alkaline conditions. Polydopamine has been used as a biocompatible and biodegradable coating for organic/inorganic materials and nanoparticles/microparticles.[27,28] The polydopamine coating thickness can be controlled by varying the concentration of dopamine, and reaction time and redox properties of catechol groups in polydopamine can be used for functionalization with amine/thiol-terminated biomolecules through a Schiff base or the Michael addition reaction.

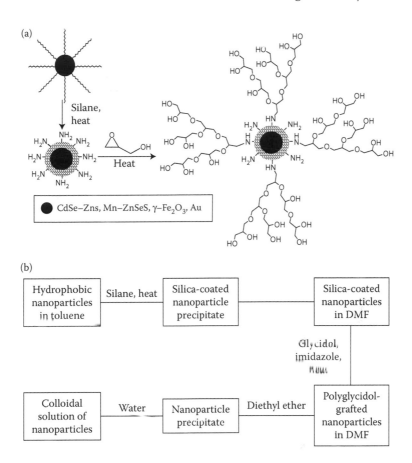

FIGURE 4.9 Schematic representation of hyperbranched polyglycerol grafting around primary amine-terminated colloidal nanoparticles with the chemistry involved (a) and experimental steps (b). (Reprinted with permission from Panja, P. et al. 2017. Hyperbranched polyglycerol grafting on the surface of silica-coated nanoparticles for high colloidal stability and low nonspecific interaction. *ACS Sustainable Chemistry & Engineering*, 5, 4879–4889. Copyright 2017 American Chemical Society.)

4.8.1 Polydopamine-Coated Polymeric Nanoparticles

Take 2–5 mL polymeric nanoparticle dispersion in water or phosphate buffer solution.[27–29] In a separate vial, dissolve 1–5 mg dopamine in 0.1–0.5 mL water and add to the colloidal nanoparticle solution. Stir the solution mixture for 1–12 h. Next, collect the particles by centrifuge and redisperse them in fresh water. Particles can be further purified by dialysis for 24 h using nitrocellulose membranes (molecular weight cutoff: 12,000–14,000 Da).

4.8.2 Polydopamine-Coated Inorganic Nanoparticles

This approach is applicable to most hydrophilic nanoparticles.[27,30,31] Take 2–5 mL aqueous solution of colloidal nanoparticles and mix with 2–3 mL tris buffer solution of pH 8.5. Next, add a small amount of dopamine into the dispersion, to make the final concentration of dopamine 0.1–1 mg/mL. Shake the solution for another 30 min to make the dopamine self-polymerize on the surface of the nanoparticles. Centrifuge the nanoparticles and wash with water twice, then redisperse the particles in water for further use. The thickness of polydopamine will vary depending on the concentration of dopamine used.

4.8.3 Polydopamine-Coated Reduced Graphene Oxide

Take 100 mL colloidal graphene oxide (0.5 mg/mL) in a tris buffer solution of pH 8.5.[32] Add 50 mg of dopamine hydrochloride. Stir the reaction mixture vigorously at 60°C for 24 h. Next, filter the solution with a 0.2-μm membrane, centrifuge to isolate the particles and wash with fresh water. Then, redisperse the black powder in fresh water.

REFERENCES

1. Jana, N. R., Earhart, C. and Ying, J. Y. 2007. Synthesis of water-soluble and functionalized nanoparticles by silica coating. *Chemistry of Materials*, 19, 5074–5082.
2. Das, P. and Jana, N. R. 2016. Length-controlled synthesis of calcium phosphate nanorod and nanowire and application in intracellular protein delivery. *ACS Applied Materials & Interfaces*, 8, 8710–8720.
3. Sinha, A., Basiruddin, S. K., Chakraborty, A. and Jana, N. R. 2015. β-cyclodextrin functionalized magnetic mesoporous silica colloid for cholesterol separation. *ACS Applied Materials & Interfaces*, 7, 1340–1347.

4. Chakraborty, A., Dalal, C. and Jana, N. R. 2018. Colloidal nanobioconjugate with complementary surface chemistry for cellular and subcellular targeting. *Langmuir*, 34, 13461–13471.

5. Saha, A., Basiruddin, S. K., Maity, A. R. and Jana, N. R. 2013. Synthesis of nanobioconjugates with a controlled average number of biomolecules between 1 and 100 per nanoparticle and observation of multivalency dependent interaction with proteins and cells. *Langmuir*, 29, 13917–13924.

6. Chakraborty, A. and Jana, N. R. 2015. Clathrin to lipid raft-endocytosis via controlled surface chemistry and efficient perinuclear targeting of nanoparticle. *The Journal of Physical Chemistry Letters*, 6, 3688–3697.

7. Dalal, C., Saha, A. and Jana, N. R. 2016. Nanoparticle multivalency directed shifting of cellular uptake mechanism. *The Journal of Physical Chemistry C*, 120, 6778–6786.

8. Dalal, C. and Jana, N. R. 2017. Multivalency effect of TAT-peptide-functionalized nanoparticle in cellular endocytosis and subcellular trafficking. *The Journal of Physical Chemistry B*, 121, 2942–2951.

9. Debnath, K., Pradhan, N., Singh, B. K., Jana, N. R. and Jana, N. R. 2017. Poly(trehalose) nanoparticles prevent amyloid aggregation and suppress polyglutamine aggregation in a Huntington's disease model mouse. *ACS Applied Materials & Interfaces*, 9, 24126–24139.

10. Saha, A., Basiruddin, S. K., Sarkar, R., Pradhan, N. and Jana, N. R. 2009. Functionalized plasmonic–fluorescent nanoparticles for imaging and detection. *The Journal of Physical Chemistry C*, 113, 18492–18498.

11. Basiruddin, S. K., Saha, A., Sarkar, R., Majumder, M. and Jana, N. R. 2010. Highly fluorescent magnetic quantum dot probe with superior colloidal stability. *Nanoscale*, 2, 2561–2564.

12. Basiruddin, S. K., Saha, A., Pradhan, N. and Jana, N. R. 2010. Functionalized gold nanorod solution via reverse micelle based polyacrylate coating. *Langmuir*, 26, 7475–7481.

13. Saha, A., Basiruddin, S. K., Ray, S. C., Roy, S. S. and Jana, N. R. 2010. Functionalized graphene and graphene oxide solution via polyacrylate coating. *Nanoscale*, 2, 2777–2782.

14. Pellegrino, T., Manna, L., Kudera, S., Liedl, T., Koktysh, D., Rogach, A. L., Keller, S., Radler, J., Natile, G. and Parak, W. J. 2004. Hydrophobic nanocrystals coated with an amphiphilic polymer shell: A general route to water soluble nanocrystals. *Nano Letters*, 4, 703–707.

15. Palmal, S., Basiruddin, S. K., Maity, A. R., Ray, S. C. and Jana, N. R. 2013. Thiol-directed synthesis of highly fluorescent gold clusters and their conversion into stable imaging nanoprobes. *Chemistry-A European Journal*, 19, 943–949.

16. Maity, A. R., Palmal, S., Basiruddin, S. K., Karan, N. S., Sarkar, S., Pradhan, N. and Jana, N. R. 2013. Doped semiconductor nanocrystal based fluorescent cellular imaging probes. *Nanoscale*, 5, 5506–5513.

17. Bhunia, S. K., Saha, A., Maity, A. R., Ray, S. C. and Jana, N. R. 2013. Carbon nanoparticle-based fluorescent bioimaging probes. *Scientific Reports*, 3, 1473.

18. Ali, H., Bhunia, S. K., Dalal, C. and Jana, N. R. 2016. Red fluorescent carbon nanoparticle-based cell imaging probe. *ACS Applied Materials & Interfaces*, 8, 9305–9313.

19. Debnath, K., Mandal, K. and Jana, N. R. 2016. Phase transfer and surface functionalization of hydrophobic nanoparticle using amphiphilic poly(amino acid). *Langmuir*, 32, 2798–2807.

20. Bhunia, S. K. and Jana, N. R. 2011. Peptide-functionalized colloidal graphene via interdigited bilayer coating and fluorescence turn-on detection of enzyme. *ACS Applied Materials & Interfaces*, 3, 3335–3341.

21. Nandanan, E., Jana, N. R. and Ying, J. Y. 2008. Functionalization of gold nanospheres and nanorods by chitosan oligosaccharide derivatives. *Advanced Materials*, 20, 2068–2073.

22. Jana, N. R., Erathodiyil, N., Jiang, J. and Ying, J. Y. 2010. Cysteine-functionalized polyaspartic acid: A polymer for coating and bioconjugation of nanoparticles and quantum dots. *Langmuir*, 26, 6503–6507.

23. Jana, N. R. and Peng, X. 2003. Single-phase and gram-scale routes toward nearly monodisperse Au and other noble metal nanocrystals. *Journal of the American Chemical Society*, 125, 14280–14281.

24. Jana, N. R., Patra, P. K., Saha, A., Basiruddin, S. K. and Pradhan, N. 2009. Imidazole based biocompatible polymer coating in deriving <25 nm functional nanoparticle probe for cellular imaging and detection. *The Journal of Physical Chemistry C*, 113, 21484–21492.

25. Palmal, S., Jana, N. R. and Jana, N. R. 2014. Inhibition of amyloid fibril growth by nanoparticle coated with histidine-based polymer. *The Journal of Physical Chemistry C*, 118, 21630–21638.

26. Panja, P., Das, P., Mandal, K. and Jana, N. R. 2017. Hyperbranched polyglycerol grafting on the surface of silica-coated nanoparticles for high colloidal stability and low nonspecific interaction. *ACS Sustainable Chemistry & Engineering*, 5, 4879–4889.

27. Mandal, K. and Jana, N. R. 2018. Galactose-functionalized, colloidal-fluorescent nanoparticle from aggregation-induced emission active molecule via polydopamine coating for cancer cell targeting. *ACS Applied Nano Materials*, 1, 3531–3540.
28. Liu, M., Zeng, G., Wang, K., Wan, Q., Tao, L., Zhang, X. and Wei, Y. 2016. Recent developments in polydopamine: An emerging soft matter for surface modification and biomedical applications. *Nanoscale*, 8, 16819–16840.
29. Park, J., Brust, T. F., Lee, H. J., Lee, S. C., Watts, V. J. and Yeo, Y. 2014. Polydopamine-based simple and versatile surface modification of polymeric nano drug carriers. *ACS Nano*, 8, 3347–3356.
30. Wang, S., Zhao, X., Wang, S., Qian, J. and He, S. 2016. Biologically inspired polydopamine capped gold nanorods for drug delivery and light-mediated cancer therapy. *ACS Applied Materials Interfaces*, 8, 24368–24384.
31. Zheng, R., Wang, S., Tian, Y., Jiang, X., Fu, D., Shen, S. and Yang, W. 2015. Polydopamine-coated magnetic composite particles with an enhanced photothermal effect. *ACS Applied Materials and Interfaces*, 7, 15876–15884.
32. Huang, N., Zhang, S., Yang, L., Liu, M., Li, H., Zhang, Y. and Yao, S. 2015. Multifunctional electrochemical platforms based on the Michael Addition/Schiff base reaction of polydopamine modified reduced graphene oxide: Construction and application. *ACS Applied Materials and Interfaces*, 7, 17935–17946.

CHAPTER 5

Selected Conjugation Chemistries for Making Chemically or Biochemically Functionalized Colloidal Nanoparticles

5.1 INTRODUCTION

Conjugation chemistry is critical for covalent linking of nanoparticles with chemicals/biochemicals of interest. This chapter summarizes various conjugation chemistries and standard procedures for covalent conjugation of chemicals/biochemicals with coated nanoparticles. These conjugation chemistries can be adapted in transforming coated nanoparticles to functional nanoparticles. Conjugation chemistries are selected based on simplicity, effectiveness and general applicability to different nanoparticles. Conjugation methods include reaction of primary amine-terminated nanoparticles with N-hydroxysuccinamide (NHS)-based reagents, EDC (1-ethyl-3-(3-(dimethylamino)propyl)carbodiimide chloride) coupling between primary amine-terminated nanoparticles and carboxyl-bearing molecules, glutaraldehyde-based conjugation between primary amine-terminated nanoparticles and primary amine-bearing molecules, di-NHS-based conjugation between primary amine-terminated nanoparticles and primary amine-terminated molecules, 4-maleimidobutyric acid N-hydroxy succinimide ester (MAL-but-NHS)-based conjugation between primary amine-terminated nanoparticles and thiol-bearing molecules, Schiff-base reaction between primary amine-terminated nanoparticles and aldehyde-bearing molecules, epoxide-based reactions, disuccinimidyl carbonate-based conjugation, isothiocyatate-based conjugation, Michael addition of thiolated molecules with polydopamine-coated nanoparticles, polymer coating using designed monomers, chloroformylate-based reactions and thiol-based approaches. In particular, standard conjugation procedures are described that produce a variety of nanobioconjugates. The described conjugation approaches offer general guidelines to prepare different nanobioconjugates and can be adapted to a new generation of nanoparticles of interest.

5.2 REACTION OF PRIMARY AMINE-TERMINATED NANOPARTICLES WITH N-HYDROXYSUCCINAMIDE-BASED REAGENTS

Many NHS-based reagents are commercially available, and some NHS reagents can be made by adapting known organic chemistry principles. Here, conjugation methods with two such NHS-based reagents (biotin-NHS, folate-NHS) are described that are most frequently used in deriving biotin/folate-functionalized nanoparticles.

5.2.1 Biotinylated Nanoparticles

Dissolve 1–2 mg biotin-NHS in 0.1–0.2 mL anhydrous dimethylformamide.[1-3] In a separate vial, take 1–2 mL of primary amine-terminated nanoparticle solution and mix it with 1–2 mL borate buffer solution of pH 9. Next, mix with 50–100 μL of biotin-NHS solution and stir overnight under room temperature. After that, dialyze the solutions (using a membrane with molecular weight cutoff 12,000–14,000 Da) against water to remove excess reagents. Finally, preserve the solutions at 4°C.

5.2.2 Folate-Functionalized Nanoparticles

First, prepare the NHS derivative of folic acid (folate-NHS) by using N,N'-dicyclohexylcarbodiimide (DCC)-based coupling chemistry.[2,4,5] Typically, dissolve 0.3 g folic acid and 0.15 mL triethylamine in 10 mL freshly distilled dimethyl sulfoxide (DMSO), then add 0.14 g DCC. Stir the solution for 1 h at room temperature and add 0.12 g NHS. Stir the mixture overnight in the dark at room temperature, then centrifuge to remove the insoluble hyproduct. Next, precipitate the product by adding minimum diethyl ether, collect the precipitate by centrifuge, wash with dry tetrahydrophuran and dry the solid under vacuum. Preserve the resultant folate-NHS at 4°C.

Next, take 2 mL aqueous solution of primary amine-terminated nanoparticles (0.1–1.0 mg/mL) in a vial and mix with 0.2 mL borate/phosphate buffer of pH 9.0. Next, add 0.2 mL dimethylformamide solution of folate-NHS (8–72 mg/mL) and stir overnight. Next, dialyze the solution first against normal water, then against basic water and finally against normal water. Finally, preserve the folate-conjugated nanoparticles at 4°C.

5.3 1-ETHYL-3-(3-(DIMETHYLAMINO)PROPYL CARBODIIMIDE CHLORIDE) COUPLING BETWEEN PRIMARY AMINE-TERMINATED NANOPARTICLES AND CARBOXYL-BEARING MOLECULES

1-ethyl-3-(3-(dimethylamino)propyl carbodiimide (EDC) is the most commonly used reagent that covalently links carboxylic acid groups with primary amines, and the reaction can be performed in water. Here, we describe the EDC coupling method between primary amine-terminated nanoparticles and carboxylic acid-bearing biomolecules for deriving various nanobioconjugates.

5.3.1 Proline-Conjugated Nanoparticles

Mix 0.1 mL aqueous solution of L-proline (0.01 M) with 0.1 mL 2-(N-morpholino)ethanesulfonic acid (MES) buffer of pH 4.5.[6] Next, add 0.1 mL of freshly prepared solution of N-hydroxy-succinamide (0.05 M) and 0.1 mL of EDC (0.05 M). In a separate vial, mix 0.5 mL of primary amine-terminated nanoparticles with 0.1 mL borate buffer of pH 9 and immediately add to the earlier solution mixture. Keep the mixture overnight under stirring conditions and finally dialyze the solution for 1 day against water using a dialysis membrane (molecular weight cutoff: 12,000–14,000 Da) to remove unbound reagents.

5.3.2 Trehalose-Conjugated Nanoparticles

First, prepare carboxylated trehalose as described below.[7] Dissolve 175 mg trehalose in 10 mL dry dimethylformamide. In a separate vial, dissolve 48 mg of succinic anhydride in 1 mL of dry dimethylformamide, and mix this solution with the trehalose solution. Next, add 1 mL of dimethylformamide-tetrahydrophuran mixture (100:7 volume ratio), and continue the reaction for 12 h at 80°C under a nitrogen atmosphere. Precipitate the resultant trehalose monocarboxylate using a diethyl ether/acetone mixture (70:30 volume ratio) and dry in air.

Dissolve 50 mg trehalose monocarboxylic acid in 0.5 mL MES (2-(N-morpholino) ethanesulfonic acid) buffer of pH 5. Next, mix 38 mg of EDC and 38 mg of N-hydroxy-succinamide under stirring. Immediately add 0.5 mL aqueous solution of primary amine-terminated nanoparticles (5 mg/mL). Continue the reaction for 12 h at room temperature, then precipitate the trehalose-functionalized nanoparticles by adding ethanol and purify by dialysis using a dialysis membrane (molecular weight cutoff: 12,000–14,000 Da).

5.3.3 β-Cyclodextrin-Conjugated Magnetic Mesoporous Silica Nanoparticles

First, prepare carboxylated β-cyclodextrin as mentioned below. Dissolve 1.12 g β-cyclodextrin and 0.15 g succinic anhydride in 8 mL dimethylformamide and place it in a three-necked flask equipped with mechanical stirrer, thermometer and condenser.[8] Next, add 0.14 mL triethylamine and purge the solution with nitrogen. Increase the temperature to 80°C and hold at this temperature for another 12 h. Next, cool it to room temperature and add excess chloroform for precipitation. Collect the precipitated β-cyclodextrin by centrifuge, wash with acetone and dry in vacuum.

Next, prepare 10 mL primary amine-terminated magnetic mesoporous silica nanoparticle solution with a particle concentration of 8 mg/mL (see Section 4.2.5). Next, add 250 mg of carboxylated β-cyclodextrin and adjust the pH to 5.0 by adding MES (2-(N-morpholino) ethanesulfonic acid) buffer solution of pH 5. Next, add 38 mg EDC and 38 mg N-hydroxy-succinamide to this solution and stir for 24 h. Separate the particles from the solution by centrifuge and wash four times by water and dry under vacuum or disperse in water.

5.3.4 Phenylboronic Acid-Functionalized Magnetic Mesoporous Silica Nanoparticles

Dissolve 40 mg 3-carboxyphenylboronic acid in an MES (2-(N-morpholino)ethanesulfonic acid) buffer of pH 7.0.[9] Next, add 50 mg EDC and 50 mg NHS (N-hydroxysuccinimide) to this solution under stirring conditions. After 30 min, mix this solution with 10 mL of magnetic mesoporous silica nanoparticle solution (see Section 4.2.5) and stir the whole solution for another 24 h. Next, separate the particles from the solution by centrifuge and wash with water and dimethylformamide 4–6 times. Finally, disperse the particles in 10 mL water.

5.3.5 Curcumin-Functionalized Au Nanoparticles

First, prepare the monocarboxylic acid derivative of curcumin as described below.[10] Mix 500 mg curcumin and 28 mg 4-(dimethylamino) pyridine in 25 mL tetrahydrofuran. Next, add 0.33 mL triethylamine with the change of color of the solution from yellow to deep brown. In a separate vial, dissolve 171 mg glutaric anhydride in 1.25 mL tetrahydrophuran and add dropwise to the curcumin solution with stirring. Next, heat the solution to reflux under an argon atmosphere for 24 h. Next, remove the tetrahydrophuran by using a rotary evaporator, redissolve the dried sample in 10 mL ethyl acetate and then add dilute aqueous HCl with vigorous shaking. Collect the organic phase containing curcumin monocarboxylic acid and dry the extract. Purify the product by column chromatography, eluting with a mixture of dichloromethane and methanol (95:5, v/v). The yield of curcumin-COOH is about 45%.

Next, dissolve 4 mg curcumin monocarboxylic acid in 0.25 mL ethanol and add 0.1 mL borate buffer of pH 9. In a separate vial, dissolve 16 mg EDC in 0.25 mL ethanol and add to the curcumin solution with stirring. Next, mix 1.0 mL primary amine-terminated aqueous Au nanoparticles with the previous solution and allow 6 h under stirring. Centrifuge the resulting solution at 12,000 rpm to precipitate the gold nanoparticle–curcumin conjugate. Repeatedly wash the precipitate with chloroform and ethanol to remove unbound curcumin monocarboxylic acid. Finally, dissolve the precipitated particles in 1.0 mL distilled water and use for further study.

5.3.6 Cyclic Arg-Gly-Asp (RGD) Peptide-Conjugated, Hyperbranched Polyglycerol Grafted Nanoparticles

First, transform hyperbranched polyglycerol grafted nanoparticles to carboxylic acid-functionalized nanoparticles, as described below.[11] Dissolve 100–200 mg nanoparticles (see Section 4.7) and 100 mg of succinic anhydride in 10 mL dry dimethylformamide. Next, mix 120 mg 4-dimethylamino pyridine and 1 mL triethylamine and heat at 70°C for 24 h under an argon atmosphere. Under this condition, the hydroxy groups of polyglycerol react with succinic anhydride. Next, precipitate the nanoparticles by adding acetone and collect particles via centrifuge. In order to remove excess reagent, dissolve the product in methanol and precipitate the particles by adding acetone. Repeat this process of methanol-induced dissolution and acetone-induced precipitation three to four times. Finally, dissolve the product in distilled water and dialyze against distilled water using a dialysis membrane (molecular weight cutoff: 2000 Da) that removes excess reagents, methanol and acetone.

Next, take 2 mL of the above solution with 10 mg/mL concentration and mix with 200 μL of freshly prepared aqueous solution of EDC (20 mg/mL) and 200 μL aqueous solution of N-hydroxy susccinimide (25 mg/mL). Adjust the pH of the solution to 7.4 by adding a phosphate buffer, and stir the solution for 30 min. Next, add 500 μL aqueous solution of cyclo(Arg-Gly-Asp-d-Phe-Lys) peptide (2 mg/mL) and stir overnight. Next, remove excess reagents and unbound peptide by dialysis using a dialysis membrane (molecular weight cutoff: 2000 Da) against distilled water.

5.3.7 Vancomycin-Functionalized Nanoparticles

Take 1–2 mL primary amine-terminated nanoparticle solution, mix with 0.2 mL of aqueous solution of vancomycin (2 mg/mL) and then adjust the pH to 6.0 using MES (4-morpholineethanesulfonic acid) buffer solution.[12] In separate vials, prepare fresh aqueous solutions of EDC (0.1 M) and N-hydroxysuccinimide (0.1 M) separately, add 0.2 mL of each solution to the nanoparticle solution and incubate overnight. Next, dialyze the resulting solution overnight to remove excess reagents and preserve at 4°C.

5.4 GLUTARALDEHYDE-BASED CONJUGATION BETWEEN PRIMARY AMINE-TERMINATED NANOPARTICLES AND PRIMARY AMINE-BEARING MOLECULES

Glutaraldehyde-based conjugation is the simplest approach for covalent linking between two of the primary amine-bearing molecules. Here, we describe this approach for covalent linking between primary amine-terminated nanoparticles and primary amine-bearing molecules for deriving various nanobioconjugates.

5.4.1 Glutamine/Phenyl Alanine/Glucose-Conjugated Nanoparticles

Take 1.0 mL primary amine-terminated nanoparticle solution in a vial and mix with 0.2 mL borate buffer of pH 9.0.[6,13,14] In a separate vial, prepare 1.0 mL glutamine (or phenyl alanine or glucosamine) solution (0.01 M) in a borate buffer of pH 9.0 and mix with 1.0 mL ethanolic solution of glutaraldehyde (0.01 M). After 15 min, add 100 μL of this mixture to 0.5 mL of nanoparticle solution. After 1 h, add $NaBH_4$ solution to reduce the imine bond formed by the reaction between the aldehyde and amine. After 1 h, precipitate the nanoparticles by adding acetone followed by high-speed centrifuge at 12,000 rpm. Finally, redissolve the particles in 0.5 mL fresh water and dialyze the solution for 24 h against water using a dialysis membrane (molecular weight cutoff: 12,000–14,000 Da) to remove unbound reagents.

5.4.2 Octlyamine/Oleylamine-Conjugated Nanoparticles

Prepare an equimolar mixture of glutaraldehyde and octylamine/oleylamine (each with 0.01 M), and mix 0.1 mL of this into the 1–2 mL colloidal solution of primary amine-terminated nanoparticles.[13,15] After 1 h, add 0.1 mL of $NaBH_4$ solution (0.01 M) to reduce the imine bond, and continue the reaction for another 4 h. Next, dialyze the solution (with a molecular weight cutoff membrane of 12,000–14,000 Da) to remove the free reactants. In addition, remove the unreacted octylamine/oleylamine via repeated extraction with chloroform.

5.5 NHS-PEG-NHS (BIS[2-(N-SUCCINIMIDYL-SUCCINYLAMINO)-ETHYL] PEG-3000)-BASED CONJUGATION BETWEEN PRIMARY AMINE-TERMINATED NANOPARTICLES AND PRIMARY AMINE-BEARING MOLECULES

NHS-PEG-NHS-based conjugation is similar to glutataldehyde-based conjugation for covalent linking between two of the primary amine-bearing molecules. One significant difference over glutaraldehyde-based conjugation is that this method does not require a borohydride reduction step. Here, we describe this approach for covalent linking between primary amine-terminated nanoparticles and primary amine-bearing molecules for deriving various nanobioconjugates.

5.5.1 Cyclodextrin-Conjugated Nanoparticles

Use 3-amino-3-deoxy-β-cyclodextrin hydrate (β-CD-NH$_2$) for conjugation with primary amine-terminated nanoparticles.[16] Mix 100 μL dimethylformamide solution of NHS-PEG-NHS (4.6 mg/mL) with 100 μL borate buffer solution (pH 9.0) of β-CD-NH$_2$ (11.3 mg/mL). Shake the mixture for 2–3 min and then add the whole solution to 1–2 mL nanoparticle solution and stir for 4–5 h at room temperature. Next, dialyze the solution against fresh water for 1 day and use it as the stock solution.

5.5.2 Aptamer/Antibody Conjugated Nanoparticles

Dissolve 10 mg NHS-PEG-NHS in 0.1 mL dry DMSO. In a separate vial, prepare 1–2 mL primary amine-terminated nanoparticle solution in a borate buffer of pH 7.0 and mix with NHS-PEG-NHS solution.[17] After 15 min, isolate the nanoparticles via high-speed centrifuge and mix them with 1–2 mL aqueous solution of aptamer/antibody. After 5–6 h, isolate the nanoparticles by centrifuge and redisperse in fresh water. Alternatively, dialyze the solution to remove free reagents. Preserve the solution at 4°C for further use.

5.6 MAL-BUT-NHS (4-MALEIMIDOBUTYRIC ACID N-HYDROXY SUCCINIMIDE ESTER)-BASED CONJUGATION BETWEEN PRIMARY AMINE-TERMINATED NANOPARTICLES AND THIOL-BEARING MOLECULES

MAL-but-NHS is most commonly used for covalent linking between primary amines and thiols. Here, we describe this conjugation method between primary amine-terminated nanoparticles and thiol-bearing biomolecules for deriving various nanobioconjugates.

5.6.1 Trans Activating Transcription (TAT) Peptide-Functionalized Nanoparticles

Take 1.0 mL aqueous solution of primary amine-terminated nanoparticles in a borate buffer of pH 9.[18] In a separate vial, dissolve 1 mg MAL-but-NHS in 1 mL dimethylformamide and add 30–70 μL of it. After 10 min, add 50–100 μL of TAT-peptide (CGRKKRRQRRR, MW 1499 Da, 12 mg/mL) solution and keep it under stirring conditions for the next 1–12 h. Finally, dialyze (with a molecular weight cutoff membrane of 12,000–14,000 Da) the solution against fresh water to remove excess reagents.

5.6.2 Colloidal Graphene Functionalized with Enzyme-Cleavable Peptide

Take 1 mL colloidal reduced graphene oxide (see Sections 4.4.4 and 4.3.8) and mix with 200 μL phosphate buffer (pH~8.5).[19] Next, add 25 μL solution of MAL-but-NHS (1.4 mg dissolved in 100 μL dimethylformmide). After 15 min, add 100 μL peptide (FITC-Gly-Gly-Trp-Gly-Cys, MW 792.8) solution (1.3 mg dissolved in 200 μL dimethylformamide). Then, keep the mixture at 4°C for 6 h followed by overnight dialysis to remove excess peptides and reagents.

5.7 SCHIFF BASE REACTION BETWEEN PRIMARY AMINE-TERMINATED NANOPARTICLES AND ALDEHYDE-BEARING MOLECULES

The Schiff base reaction is used for covalent linking between two molecules with primary amines and aldehydes. Here, we describe this conjugation method between primary amine-terminated nanoparticles and aldehyde-bearing biomolecules for deriving various nanobioconjugates.

5.7.1 Glucose/Galactose/Dextan-Functionalized Nanoparticles by Cyanoborohydride (Na(CN)BH$_3$)-Based Conjugation Chemistry

This approach is not applicable to Ag/Ag nanoparticles, as cyanide can react and dissolve the Au/Ag particles.[20] However, this method can be used for iron oxide nanoparticles, quantum dots and other nanoparticles (Figure 5.1). Take 2 mL primary amine-terminated nanoparticle solution (1–5 mg/mL) in a 5-mL reaction flask and add 1–2 mL borate buffer solution of pH 9. Next, add 10–12 mg solid maltose/lactose/dextran and thoroughly mix, followed by the addition of 50–60 mg solid $Na(CN)BH_3$, and keep the solution under magnetic stirring conditions overnight. Next, purify the carbohydrate-functionalized nanoparticle solution from excess reagents either by dialysis or via microcentrifuge filtration. Typically, dialyze the functionalized nanoparticles using a 12–14 kDa molecular weight cutoff membrane against distilled water. Remove the high-molecular-weight free dextran from the dextran-functionalized nanoparticles via microcentrifuge filtration (molecular weight cutoff: 30 kDa).

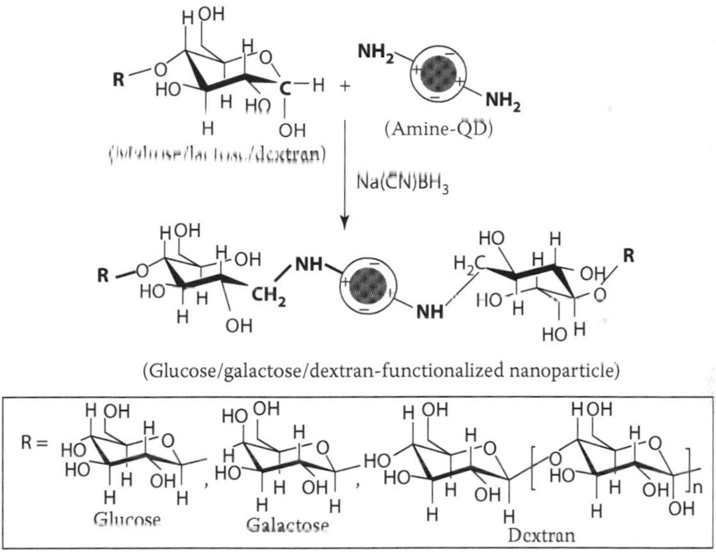

FIGURE 5.1 Schematic representation of carbohydrate-functionalized nanoparticle via cyanoborohydride-based conjugation chemistry. (Basiruddin, S. K. et al. 2012. Glucose/galactose/dextran-functionalized quantum dots, iron oxide and doped semiconductor nanoparticles with 100 nm hydrodynamic diameter. *RSC Advances*, 2, 11915–11921. Reproduced by permission of The Royal Society of Chemistry.)

5.7.2 Triphenylphosphonium-Functionalized Nanoparticles

In a 2-mL vial, take 500 μL of primary amine-terminated nanoparticle solution and mix with 200 μL aqueous solution of formylmethyl-triphenylphosphonium chloride (10 mM).[21] Stir the solution for 1 h, add 2 mg of sodium cyanoborohydride and continue stirring for another 6 h. Next, dialyze the solution overnight by using a molecular weight cutoff membrane of 12,000–14,000 Da to remove unreacted reagents.

5.7.3 Green Tea Polyphenol-Conjugated Polymer Micelle via Reaction of Primary Amine-Terminated Micelles with Oxidized Polyphenol

A schematic representation of this synthesis approach is shown in Figure 5.2.[22] Dissolve 250 mg of polysuccinimide in 20 mL dry dimethylformamide (see Section 3.15). Mix with 135 mg octadecylamine and heat at 80°C for 24 h under an inert atmosphere. Cool the solution to room temperature, mix it with 170 μL of ethylenediamine and heat it further at 80°C for 24 h under an inert atmosphere. In a separate vial, prepare green tea polyphenol solution in dimethylformamide with a concentration of 10 mg/mL. Mix 0.5–1.0 mL of dimethylformmide solution of polymer with 0.1–0.2 mL of green tea polyphenol solution. Next, add this mixture dropwise to 5–10 mL of water under vigorous stirring conditions, and continue stirring for 6 h. Next, remove dimethylformamide and free polyphenol via dialysis (molecular weight cutoff: 12,000–14,000 Da) of the whole colloidal solution against distilled water. Use this as a stock solution.

5.7.4 Polyethylene Glycol-Functionalized Nanoparticles

Take 1–2 mL primary amine-terminated nanoparticle solution in a borate buffer of pH 9.5.[23] In a separate vial, dissolve 10 mg methoxypoly(ethylene glycol) butylaldehyde in 0.2 mL water and mix it with nanoparticle solution. After 1 h, add 0.1 mL freshly prepared aqueous borohydride solution (0.1 M) to reduce the unstable imine bond. After 30 min, dialyze this solution overnight to remove any unreacted reagents.

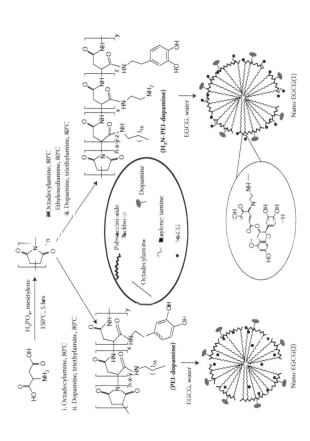

FIGURE 5.2   Schematic representation of green tea polyphenol-conjugated polymer micelle via reaction of primary amine-terminated micelle with oxidized polyphenol. (Reprinted with permission from Debnath, K. 2016. Efficient inhibition of protein aggregation, disintegration of aggregates, and lowering of cytotoxicity by green tea polyphenolbased self-assembled polymer nanoparticles. *ACS Applied Materials & Interfaces,* 8, 20309–20318. Copyright 2016 American Chemical Society.)

5.8 EPOXIDE-BASED REACTIONS

Epoxide reacts with primary amines or alcohols and offers covalent linking between two molecules with epoxide and primary amine/ alcohol. Epoxide groups are highly abundant in freshly prepared graphene oxide (prepared via Hummer's method) and can be reacted with primary amine/alcohol-terminated molecules to derive various functional graphenes/graphene oxides.

5.8.1 Cyclodextrin-Functionalized Colloidal Graphene Oxide/Reduced Graphene Oxide

Prepare fresh colloidal graphene oxide with a concentration of 1 mg/mL.[24] (see Section 3.14) In a separate vial, prepare cyclodextrin solution with a concentration of 50 mg/mL. Next, mix 10 mL of graphene oxide solution with 10 mL of cyclodextrin solution, followed by the addition of 200 μL of NH_3 solution (25 wt %) under stirring conditions. Next, add 200 μL of hydrazine (98%), raise the temperature to 70°C–80°C and maintain for 4 h. The color of the solution gradually turns black, along with the appearance of partial precipitation. Next, stop the reaction and add 0.5 mL of NaCl solution (~20 mg/mL) to precipitate the particles. Wash the precipitate several times with distilled water and finally disperse in water for further use.

5.8.2 Dextran-Functionalized Colloidal Graphene Oxide/Reduced Graphene Oxide

Take 2.0 mL freshly prepared colloidal solution of graphene oxide (1 mg/mL)[25] (see Section 3.14). Next, mix 150 mg of solid dextran (MW 6000) and stir to dissolve. Next, add 10 μL of hydrazine monohydrate and 5 μL of ammonia solution (25%) and raise the temperature of the solution to 80°C with constant stirring. Under this condition, graphene oxide is reduced to graphene and covalently linked with dextran. Next, add solid NaCl until the complete precipitation of graphene. Wash the precipitated particles with water and finally disperse in water via sonication.

5.9 DISUCCINIMIDYL CARBONATE-BASED CONJUGATION

Disuccinimidyl carbonate (DSC) is a convenient reagent for preparing N-succinimidyl esters. Here, we describe this conjugation method for the preparation of activated dextran that can be used for linking with primary amine-terminated nanoparticles.

5.9.1 Preparation of Dextran-Functionalized Nanoparticles

First, prepare activated dextran by the method described below. Prepare dextran solution (30 mM) in dimethylsulfoxide (DMSO).[26] Keep the solution under stirring conditions, add an equivalent amount of DMSO solution of DSC and DMSO solution of 4-dimethylamino pyridine, and keep the solution under stirring conditions for 6 h. Precipitate the dextran by a slow addition of acetone and wash with acetone. Repeat this precipitation-redispersion several times. Dry the activated dextran powder and store at 4°C.

Next, take 1–5 mL primary amine-terminated nanoparticle solution (1–10 mg/mL) in a phosphate buffer (pH 7.5) solution. Next, add activated dextran powder in the nanoparticle solution so that the dextran concentration is about 6 mM. Shake the solution overnight at 4°C. Finally, separate the unreacted dextran through size exclusion chromatography using a Sephadex (G25) column or by overnight dialysis (using a 12–14 kDa molecular weight cutoff membrane and against deionized water) for dextran 1 K and dextran 6 K.

5.10 ISOTHIOCYATATE-BASED CONJUGATION

Isothiocyatate reacts with primary amine groups and covalently links between two molecules with isothioyanate and primary amines. Here, we describe this conjugation method between primary amine-terminated nanoparticles and the isothiocyanate derivative of fluorescent dye for deriving fluorescent dye-conjugated nanoparticles.

5.10.1 Rhodamine B-Conjugated Hydroxyapatatite Nanorods

Take 1 mL of aqueous solution of primary amine-terminated hydroxyapatatite nanorods (4 mg/mL) in a vial, and mix with

0.5–1.0 mL bicarbonate buffer of pH 9.0.[27] Next, add 10 μL of rhodamine B isothiocyanate (0.1 mM in dry dimethylformamide) solution and stir the solution overnight. Next, remove the unbound rhodamine B isothiocyanate by dialysis using a dialysis membrane (molecular weight cutoff: 12,000–14,000 Da).

5.11 REACTION OF THIOL/PRIMARY AMINE-BEARING MOLECULES WITH POLYDOPAMINE-COATED NANOPARTICLES

The polydopamine surface can react with thiol and primary amine groups due to the presence of a reactive unsaturated double bond. Here, we describe this reaction between polydopamine-coated nanoparticles and thiol/primary amine-bearing biomolecules for deriving various nanobioconjugates.

5.11.1 Galactose-Functionalized Nanoparticles

First, prepare thiolated galactose as described below. Dissolve 9 mg lactose in 0.5 mL borate buffer solution of pH 9.0 and mix with 2 mg of cysteamine (dissolve in 0.5 mL of borate buffer of pH 9) and stir for 15 min.[28] Next, add 50 mg of solid NaCNBH$_3$, stir overnight and then precipitate the thiolated galactose by adding excess acetone followed by repeated washing. Finally, dissolve the precipitate in 1.0 mL water.

Next, disperse the freshly prepared polydopamine-coated nanoparticles (1–2 mg/mL) in a phosphate buffer solution of pH 8.5, then add 100 μL thiolated galactose solution. Continue stirring for 3 h and purify the functionalized nanoparticles by centrifugation at 12,000 rpm followed by washing three times with water, and finally disperse in 1.0 mL water.

5.11.2 Biomolecule-Functionalized Reduced Graphene Oxide by Reaction with Primary Amine-Bearing Biomolecules

Take 10 mL colloidal reduced graphene oxide solution in a phosphate buffer solution of pH 7.4.[29] In a separate vial, dissolve 0.1–1 mg primary amine-bearing biomolecules (e.g., antibody/

oligonucleotide) and mix it with the earlier solution. Stir the reaction mixture at 35°C–40°C for 24 h. After the reaction, isolate the particles by centrifuge, wash thoroughly with water and redisperse in fresh water for further use.

5.12 PHENYLBORONIC ACID-BASED CONJUGATION CHEMISTRY

Phenylboronic acid reacts with the 1, 2 diol groups, and this reaction can be used to link between two molecules bearing phenylboronic acid and 1, 2 diol. This approach can be used to prepare nanobioconjugates.

5.12.1 Vitamin C-Conjugated Polymer Micelles

A schematic representation of this approach is shown in Figure 5.3.[30] Dissolve 250 mg of polysuccinimide (see Section 3.15) in 20 mL dry dimethylformamide. Mix with 135 mg octadecylamine and heat at 80°C for 24 h under an inert atmosphere. Cool the solution to room temperature and mix it with 170 μL of ethylenediamine and heat it further at 80°C for 24 h under an inert atmosphere. Next, add 150 mg of 4-formylphenylboronic acid and stir for 12 h at room temperature.

Next, add 1.0 mL of the above solution to 50 mL of water under vigorous stirring conditions. Next, add 5.0 mL of aqueous solution of vitamin C (0.1 M) dropwise, followed by adding base (NaOH solution) to maintain pH 7.4. Stir the solution for 3 h, then precipitate the resultant particles by adding acetone and collect by centrifuging at 12 000 rpm. Wash the precipitate with methanol and dissolve in water with a concentration of ~2 mg/mL.

5.13 POLYMER COATING USING DESIGNED MONOMERS

Acryl monomers can be derived from the desired biomolecules and then can be used as one of the monomers during the polyacrylate coating described earlier (see Section 4.3). This approach can be used to derive various nanobioconjugates.

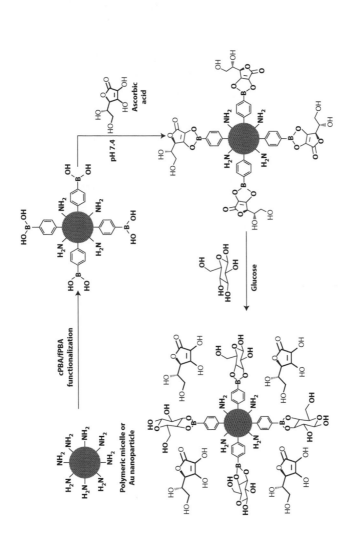

FIGURE 5.3 Schematic representation of making vitamin C-conjugated polymer nanoparticles and glucose-induced release option. (Reprinted with permission from Chakraborty, A. and Jana, N. R. 2017. Vitamin C-conjugated nanoparticle protects cells from oxidative stress at low doses but induces oxidative stress and cell death at high doses. *ACS Applied Materials & Interfaces*, 9, 41807–41817. Copyright 2017 American Chemical Society.)

5.13.1 Trehalose-Functionalized Nanoparticles via Polyacrylate Coating

Dissolve 378 mg of trehalose in 12 mL dry dimethylformamide and mix it with 836 μL of dry triethylamine under an argon atmosphere.[7] Then, add 96 μL of dry crotonoyl chloride and continue the reaction for 6 h at 55°C under an argon atmosphere. Next, cool the solution to room temperature and add excess diethyl ether to precipitate the product. Dry the solid and use it as one of the acryl monomers during polyacrylate coating, as described in Section 4.3.

5.14 CHLOROFORMYLATE-BASED REACTIONS

The chloroformyl group is reactive to primary alcohol/amine, and some chloroformyl reagents are commercially available. Nanobioconjugates can be derived from primary amine-terminated nanoparticles using this approach.

5.14.1 Cholesterol-Functionalized, Hyperbranched Polyglycerol Grafted Nanoparticles

Take 10–20 mg of hyperbranched polyglycerol grafted nanoparticles (see Section 4.7) in a water–dimethylformamide mixture.[31] In a separate vial, prepare a dimethylformamide solution of cholesteryl chloroformate (0.8 mg/mL) and add 0.1 mL of it to the nanoparticle solution. Next, add 8 μL of trimethylamine and 50 μL aqueous solution of 4-(dimethylamino)pyridine (48 mg/mL). After 4 h, dialyze the solution against distilled water using a dialysis membrane (molecular weight cutoff: 2000 Da).

5.15 THIOL-BASED APPROACH

Thiolated biomolecules can be designed as capping agents for nanoparticles. Here, we describe this approach in deriving various nanobioconjugates.

5.15.1 TAT-peptide-Functionalized Quantum Dots via Ligand Exchange

Prepare reverse micelles by mixing 1 mL of Igepal (CO-520) with 9 mL of cyclohexane.[32] Next, dissolve 5–10 mg of purified

QDs in 2 mL reverse micelles. Next, add 50 μL of aqueous TAT peptides (10 mg/mL) with cysteine at one end, followed by 50 μL of aqueous tetramethylammonium hydroxide (0.1 M) to induce ligand exchange. Sonicate for 1 min, vortex for 5 min and then add 1–2 mL ethanol to precipitate the particles. Collect the particles by centrifuge. Wash the precipitate twice with the reverse micelle solution and three times with ethanol. Then, disperse in tris(hydroxymethyl)aminomethane (Tris) buffer of pH 7.0.

5.15.2 Trehalose-Conjugated Au Nanoparticles via Reduction of Gold Salt in Presence of Thiolated Trehalose

First, prepare lipoic acid-conjugated trehalose as described below.[33] Take 210 mg trehalose, 150 mg lipoic acid and 10 mg 4-(dimethylamino) pyridine in 3 mL anhydrous dimethylformamide. Stir the mixture to dissolve the reagents. Then, cool the medium at 0°C in an ice bath. Now, add 165 mg DCC (N,N'-dicyclohexylcarbodiimide) to the reaction mixture and continue stirring at 0°C for another 30 min, followed by keeping the mixture at room temperature for 24 h. Next, precipitate the product with acetone and wash with dichloromethane thrice to remove unreacted reagents and side products.

Next, prepare 3 mL of aqueous $AuCl_3$ (1 mM) solution and add 0.2 mL aqueous solution of lipoic acid-conjugated trehalose (15 mM). Next, add 24 μL of an aqueous solution of ascorbic acid (100 mM) under stirring. Thereafter, isolate the Au-trehalose nanoparticles by centrifugation at 12,000 rpm and redisperse in 3 mL of fresh water. Figure 5.4 shows a schematic representation of a trehalose-terminated Au nanoparticle.

5.15.3 Glucose-Functionalized Au/Ag and Ag-Coated Au Nanoparticles

Dissolve 26 mg glucosamine hydrochloride in 2 mL of phosphate buffer saline (PBS) buffer of pH 7.4 and mix with 14 mg iminothiolate hydrochloride.[34] Then, stir the mixture for 4 h and characterize the thiol-modified glucose by mass and NMR spectroscopy.

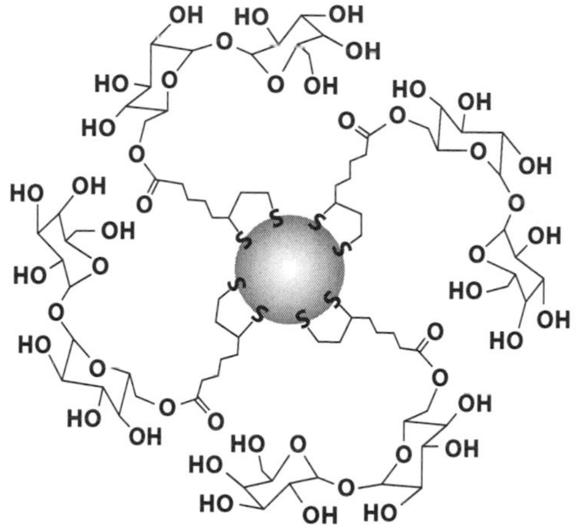

FIGURE 5.4 Schematic representation of trehalose-terminated Au nanoparticle. (Reprinted with permission from Mandal, S. et al. 2017. Trehalose-functionalized gold nanoparticle for inhibiting intracellular protein aggregation, *Langmuir*, 33, 13996–14003. Copyright 2017 American Chemical Society.)

In a separate vial, take 5 mL of Au/Ag/Ag coated Au nanoparticles (see Sections 3.2.1, 3.3 and 3.4) and mix it with 50 µL of thiol modified glucose. Stir the mixture for 4 h. Finally, dialyze or centrifuge the solution to remove any unbound reagents.

5.15.4 Biotin-Functionalized Au/Ag and Ag-Coated Au Nanoparticles

Dissolve 34 mg biotin-N-hydroxy succinimide in 1 mL dimethylformamide. In a separate vial, dissolve 11 mg cysteamine in 1 mL of water and mix it with the earlier solution.[34] Stir the mixture for 4 h and characterize the product via mass spectroscopy.

In a separate vial, take 5 mL Au/Ag nanoparticle solution (see Sections 3.2.1, 3.3 and 3.4), mix it with 50 µL thiol modified biotin solution and stir for 4 h. Finally, dialyze or centrifuge to remove any unbound reagents.

REFERENCES

1. Nandanan, E., Jana, N. R. and Ying, J. Y. 2008. Functionalization of gold nanospheres and nanorods by chitosan oligosaccharide derivatives. *Advanced Materials*, 20, 2068–2073.
2. Chakraborty, A., Maity, A. R. and Jana, N. R. 2014. Folate and biotin based bifunctional quantum dots as fluorescent cell labels. *RSC Advances*, 4, 10434–10438.
3. Ali, H., Bhunia, S. K., Dalal, C. and Jana, N. R. 2016. Red fluorescent carbon nanoparticle-based cell imaging probe. *ACS Applied Materials & Interfaces*, 8, 9305–9313.
4. Maity, A. R., Saha, A., Roy, A. and Jana, N. R. 2013. Folic acid functionalized nanoprobes for fluorescence-, dark-field-, and dual-imaging-based selective detection of cancer cells and tissue. *ChemPlusChem*, 78, 259–267.
5. Dalal, C., Saha, A. and Jana, N. R. 2016. Nanoparticle multivalency directed shifting of cellular uptake mechanism. *The Journal of Physical Chemistry C*, 120, 6778–6786.
6. Pradhan, N., Jana, N. R. and Jana, N. R. 2018. Inhibition of protein aggregation by iron oxide nanoparticles conjugated with glutamine- and proline-based osmolytes. *ACS Applied Nano Materials*, 1, 1094–1103.
7. Debnath, K., Pradhan, N., Singh, B. K., Jana, N. R. and Jana, N. R. 2017. Poly(trehalose) nanoparticles prevent amyloid aggregation and suppress polyglutamine aggregation in a Huntington's disease model mouse. *ACS Applied Materials & Interfaces*, 9, 24126–24139.
8. Sinha, A., Basiruddin, S. K., Chakraborty, A. and Jana, N. R. 2015. β-cyclodextrin functionalized magnetic mesoporous silica colloid for cholesterol separation. *ACS Applied Materials & Interfaces*, 7, 1340–1347.
9. Sinha, A., Chakraborty, A. and Jana, N. R. 2014. Dextran-gated, multifunctional mesoporous nanoparticle for glucose-responsive and targeted drug delivery. *ACS Applied Materials & Interfaces*, 6, 22183–22191.
10. Palmal, S., Maity, A. R., Singh, B. K., Basu, S., Jana, N. R. and Jana, N. R. 2014. Inhibition of amyloid fibril growth and dissolution of amyloid fibrils by curcumin–gold nanoparticles. *Chemistry-A European Journal*, 20, 6184–6191.
11. Das, P. and Jana, N. R. 2014. Highly colloidally stable hyperbranched polyglycerol grafted red fluorescent silicon nanoparticle as bioimaging probe. *ACS Applied Materials & Interfaces*, 6, 4301–4309.

12. Basiruddin, S. K., Saha, A., Sarkar, R., Majumder, M. and Jana, N. R. 2010. Highly fluorescent magnetic quantum dot probe with superior colloidal stability. *Nanoscale*, 2, 2561–2564.
13. Basiruddin, S. K., Maity, A. R., Saha, A. and Jana, N. R. 2011. Gold-nanorod-based hybrid cellular probe with multifunctional properties. *The Journal of Physical Chemistry C*, 115, 19612–19620.
14. Chakraborty, A. and Jana, N. R. 2015. Clathrin to lipid raft-endocytosis via controlled surface chemistry and efficient perinuclear targeting of nanoparticle. *The Journal of Physical Chemistry Letters*, 6, 3688–3697.
15. Mandal, S. and Jana, N. R. 2017. Quantum dot-based designed nanoprobe for imaging lipid droplet. *The Journal of Physical Chemistry C*, 121, 23727–23735.
16. Pal, S., Dalal, C. and Jana, N. R. 2017. Supramolecular host–guest chemistry-based folate/riboflavin functionalization and cancer cell labeling of nanoparticles. *ACS Omega*, 2, 8948–8958.
17. Jana, N. R. and Ying, J. Y. 2008. Synthesis of functionalized Au nanoparticles for protein detection. *Advanced Materials*, 20, 430–434.
18. Dalal, C. and Jana, N. R. 2017. Multivalency effect of TAT-peptide-functionalized nanoparticle in cellular endocytosis and subcellular trafficking. *The Journal of Physical Chemistry B*, 121, 2942–2951.
19. Basiruddin, S. K., and Jana, N. R. 2011. Peptide functionalized colloidal graphene via interdigited bilayer coating and fluorescence turn-on detection of enzyme. *ACS Applied Materials & Interfaces*, 3, 3335–3341.
20. Basiruddin, S. K., Maity, A. R. and Jana, N. R. 2012. Glucose/galactose/dextran-functionalized quantum dots, iron oxide and doped semiconductor nanoparticles with <100 nm hydrodynamic diameter. *RSC Advances*, 2, 11915–11921.
21. Chakraborty, A. and Jana, N. R. 2015. Design and synthesis of triphenylphosphonium functionalized nanoparticle probe for mitochondria targeting and imaging. *The Journal of Physical Chemistry C*, 119, 2888–2895.
22. Debnath, K., Shekhar, S., Kumar, V., Jana, N. R. and Jana, N. R. 2016. Efficient inhibition of protein aggregation, disintegration of aggregates, and lowering of cytotoxicity by green tea polyphenol-based self-assembled polymer nanoparticles. *ACS Applied Materials & Interfaces*, 8, 20309–20318.
23. Tan, S. J., Jana, N. R., Gao, S., Patra, P. K. and Ying, J. Y. 2010. Surface-ligand-dependent cellular interaction, subcellular localization, and cytotoxicity of polymer-coated quantum dots. *Chemistry of Materials*, 22, 2239–2247.

24. Mondal, A. and Jana, N. R. 2012. Fluorescent detection of cholesterol using β-cyclodextrin functionalized graphene. *Chemical Communications*, 48, 7316–7318.

25. Maity, A. R., Chakraborty, A., Mondal, A. and Jana, N. R. 2014. Carbohydrate coated, folate functionalized colloidal graphene as a nanocarrier for both hydrophobic and hydrophilic drugs. *Nanoscale*, 6, 2752–2758.

26. Earhart, C., Jana, N. R., Erathodiyil, N. and Ying, J. Y. 2008. Synthesis of carbohydrate-conjugated nanoparticles and quantum dots. *Langmuir*, 24, 6215–6219.

27. Das, P. and Jana, N. R. 2016. Length-controlled synthesis of calcium phosphate nanorod and nanowire and application in intracellular protein delivery. *ACS Applied Materials & Interfaces*, 8, 8710–8720.

28. Mandal, K. and Jana, N. R. 2018. Galactose-functionalized, colloidal-fluorescent nanoparticle from aggregation-induced emission active molecule via polydopamine coating for cancer cell targeting. *ACS Applied Nano Materials*, 1, 3531–3540.

29. Huang, N., Zhang, S., Yang, L., Liu, M., Li, H., Zhang, Y. and Yao, S. 2015. Multifunctional electrochemical platforms based on the Michael addition/Schiff base reaction of polydopamine modified reduced graphene oxide: construction and application. *ACS Applied Materials and Interfaces*, 7, 17935–17946.

30. Chakraborty, A. and Jana, N. R. 2017. Vitamin C-conjugated nanoparticle protects cells from oxidative stress at low doses but induces oxidative stress and cell death at high doses. *ACS Applied Materials & Interfaces*, 9, 41807–41817.

31. Panja, P., Das, P., Mandal, K. and Jana, N. R. 2017. Hyperbranched polyglycerol grafting on the surface of silica-coated nanoparticles for high colloidal stability and low nonspecific interaction. *ACS Sustainable Chemistry & Engineering*, 5, 4879–4889.

32. Wei, Y., Jana, N. R., Tan, S. J. and Ying, J. Y. 2009. Surface coating directed cellular delivery of TAT-functionalized quantum dots. *Bioconjugate Chemistry*, 20, 1752–1758.

33. Mandal, S., Debnath, K., Jana, N. R. and Jana, N. R. 2017. Trehalose-functionalized gold nanoparticle for inhibiting intracellular protein aggregation. *Langmuir*, 33, 13996–14003.

34. Saha, A. and Jana, N. R. 2014. Paper-based microfluidic approach for surface-enhanced Raman spectroscopy and highly reproducible detection of proteins beyond picomolar concentration. *ACS Applied Materials & Interfaces*, 7, 996–1003.

CHAPTER 6

Clinic Nanodrugs

State of the Art and Future

6.1 INTRODUCTION

The application of nanoscale materials in the area of biomedical science is a relatively new and rapidly evolving field. These applications include drug delivery nanocarriers, optical detection nanoprobes, imaging nanoprobes and nanomaterial-based theranostics.[1-3] Among them, development of clinical nanodrugs that can be used for treatment of human diseases is the most exciting and challenging. Although controversy exists regarding the definition of nanodrugs, any drug product that incorporates particles in the size range from 1 to 100 nm and exhibits key differences in comparison to bulk materials can be considered a nanodrug.[1] This chapter will summarize the advancements in this field and offer guidelines for future research in this field.

6.2 ADVANTAGES OF NANODRUG FORMULATION AND REQUIREMENTS FOR CLINICAL APPLICATIONS

Nanodrug formulations can impart several physical and biological advantages compared to conventional medicines.[1-3] These include improved solubility and pharmacokinetics, enhanced efficacy, reduced toxicity and increased selectivity. In particular, most drug molecules are organic in nature and water insoluble. In contrast, their nanoparticle forms are water soluble due to their colloidal form. In addition, interaction of the nanoparticle form of drugs with the biological environment leads to a completely different biodistribution pattern, particularly due to multivalent interaction, high cellular uptake via endocytosis and easier targeting to different organs via an enhanced permeation retention effect. Moreover, the optical/magnetic properties of nanoparticles can be used to monitor the actual delivery sites with respect to effective performance, the photothermal properties of nanoparticles can be utilized for therapeutic application and nanodrugs with multiple drugs can be used for more effective therapy.

The most significant challenge for the development of nanodrugs is the performance of new nanomaterials with respect to safety and toxicity issues.[1] A variety of nanoparticles have been used as nanocarriers in the last 50 years, and there are numerous possibilities for making nanoformulations.[3] However, only 50 nanodrugs have been approved to date, and few more are being studied in clinical trials. Thus, the properties of nanoparticles need to be well understood prior to their development.[3] In particular, the structure–properties relationships of unknown nanodrugs and their impact on biological systems need to be studied thoroughly.[1] It is certain that tissue/cells readily take up nanodrugs via passive and active mechanisms.[1] However, nanodrugs can potentially interact with many types of cells, organs and tissues on the way from the site of administration to the intended target. In addition, depending on size and physicochemical properties, nanoparticles interact with plasma proteins and immune cells.[3]

Nanoparticles are reported to interfere with organ function. Several studies have reported nanoparticle-induced oxidative stress that can lead to inflammation in the liver, lung and brain.[4,5] The free radical/oxidative activity of some nanoparticles can cause genotoxicity.[4] Some nanoparticles have been observed to cross the blood–brain barrier and cause neurotoxicity.[6] Positively charged nanoparticles have been found to cause hepatotoxicity and increase liver enzyme levels.[4] Nanoparticles can induce both immunostimulation and immune suppression.[7] Acute and chronic exposure to gold nanoparticles has been found to alter gene expression.[1] Inorganic nanoparticles are generally nonbiodegradable, persist in the environment for long periods and cause prolonged exposure of humans and animals, with unknown consequences.[8]

Preclinical testing is an essential component of nanodrug development that involves animal studies.[1] In addition, clinical trials are required to determine the performance and safety in humans.[1] Phase 1 trials involve dosing, toxicity and excretion studies; phase 2 trials involve performance and safety issues; and phase 3 trials involve randomized, extensive trials.[1] In those steps, closer collaboration between regulatory agencies is required.[1–3]

6.3 TYPES OF MATERIALS USED IN APPROVED AND INVESTIGATIONAL NANODRUGS

So far, about 50 nanodrugs are currently available for clinical trials.[1,3] Most of them are administered either orally or intravenously.[2] Various drug delivery platforms have been used in the approved nanodrugs, including liposomes, polymers, micelles, nanocrystals, metals/metal oxides and other inorganic materials and proteins.[1,2] A list of some approved nanodrugs is summarized in Table 6.1.[2] Five general conclusions can be made from these studies. First, although many types of nanocarriers have been studied, liposomal and polymeric nanoformulations are the most commonly approved nanodrugs.[1,3] Most of the inorganic and nonbiological compositions fail in the clinical stage due to biocompatibility and toxicity issues. Second, most of the approved nanodrugs use conventional drugs and show reduced toxicity rather than enhanced performance.[2] Many nanodrugs do not survive in the clinical stage because of insignificant improvements in performance.[2] Third, anticancer and antimicrobial nanodrugs are more common than other classes of drugs.[2] Other diseases for which nanodrug development is being

TABLE 6.1   List of Representative Clinic Nanodrugs with Their Chemical
Composition, Target Diseases and Specific Advantages

Nanodrug Composition	Target Disease	Advantage
Liposome-doxorubicine	Ovarian cancer	Increased delivery to the site, low toxicity
Liposome amphotericin B	Fungal infection	Low toxicity
Liposome-cytarabine	Meningitis	Increased delivery to site, low toxicity
Liposome-irinotecan	Pancreatic cancer	Increased delivery to site, low toxicity
Polymer-leuprolide acetate	Prostate cancer	Longer circulation, controlled delivery
PEG-epoetin beta	Anemia	Greater drug (aptamer) stability
PEG-coagulating factor IX	Hemophilia	Longer half life
Micelle-estradiol	Vasomotor symptoms in menopause	Controlled delivery
Nano-hydroxyapatite	Bone substitute	Mimic bone structure
Nano-paliperidone palmitate	Schizophrenia	Injectable, slow release
Albumin-paclitaxel	Breast/pancreatic cancer	Greater solubility, increased targeting
Iron-dexran	Iron deficiency	Prolonged release

considered include autoimmune diseases, anesthesia, metabolic
disorders and neurological and psychiatric diseases.[2] Fourth, most
of the nanoformulations (except liposomes) use surface coatings
composed of antifouling polymers.[1] Fifth, current nanodrug
development considers multicomponent delivery carriers more than
relatively simple monocomponent platforms. Now, we will briefly
describe each type of nanocarrier used for clinical nanodrugs.

A liposome is a spherical vesicle composed of a lipid bilayer
membrane arranged around an empty core.[1,3] Liposomes are
usually 50–150 nm in diameter, and thus are sometimes slightly
larger than conventional nanoparticles.[2] Liposomes can carry and
deliver both hydrophilic or hydrophobic drugs.[1,2] The surface of a
liposome can be conjugated with antibodies for enhanced targeting.
Liposome-based nanodrugs can circulate in the bloodstream for

an extended time, selectively accumulate at the site of a tumor or infection and offer higher drug delivery to these targets.[3] Liposomal drug nanoformulations have significantly impacted pharmacology.[1] Most liposome nanoformulations use polyethylene glycol conjugate for enhanced blood circulation. Many liposomal nanodrugs have been approved, including antifungal, anticancer and analgesic drugs.[1,2] In addition, an increasing number of liposomal nanoformulations are under trial.[1]

Many approved nanodrugs incorporate polymers.[1] Polymeric nanoparticles can be easily synthesized in large amounts, and they can be made with wide range of sizes/compositions.[3] They can be directly used as therapeutic or modifying agents for a drug. Polymeric nanodrugs are used for the treatment of relapsing–remitting multiple sclerosis, Neulasta and acute bleeding in hemophilia A.[1]

Micelles are self-assembling polymeric amphiphiles. They can be designed for high drug loading, slow drug release, controlled drug delivery and bioresponsive delivery of hydrophobic drugs.[2] The composition and structure of a micellar nanoparticle can be finely tuned to achieve the desired goal. Micellar nanoformulations are used for transdermal delivery of estradiol to vasomotor symptoms associated with menopause.[1] A micellar formulation of paclitaxel is used for the treatment of ovarian cancer.[2] Micellar formulations for other cancers are being investigated in clinical trials.[1,2] Because of the broad applicability of micellar-based nanoformulations, new products are very much expected.[1]

Protein-based nanodrugs include drugs conjugated to protein carriers.[1] Usually, natural proteins are combined with conventional drugs, and the resultant nanodrugs offer reduced toxicity.[1] For example, albumin has been used as a protein-based drug carrier.[9] Paclitaxel-conjugated 130-nm albumin is the most well known nanodrug.[9] Other protein-based nanodrugs include nanoparticle albumin bound (NAB)-docetaxel, NAB-heat shock protein inhibitor and NAB-rapamycin.[1]

Among various inorganic nanoparticles, iron oxide nanoparticles have been studied in numerous clinical trials investigating their use as contrast enhancement reagents for magnetic resonance imaging (MRI).[1] However, the majority of approved iron oxide nanodrugs are indicated as iron replacement therapies.[1] Gold nanoparticles have been studied as a drug delivery carrier for extremely toxic antitumor agents with significant cardiovascular compromise.[2] However, they are rapidly cleared by the reticuloendothelial system[2] with little clinical value. PEG-conjugated nanodrugs have been developed with a decreased clearance rate.[2] Inorganic silica nanoparticle-based nanodrugs have also been developed for cancer treatment.[1] These nanodrugs are composed of a silica core labeled with a near-infrared fluorescent dye, a targeting ligand and an antifouling polymer layer.

6.4 CONCLUSIONS

Approaches for nanodrug formulations and associated studies have advanced significantly during the last 10–15 years.[2,3] Many of the investigated nanodrugs have received approval from regulatory bodies, and many of them are in clinical trials.[2] Although most of the currently approved nanodrugs are based on conventional drugs, future nanodrugs are expected to have a broader range.[1] While there are many challenges in nanodrug development, some of them need to be given special attention. These include chemical composition, biodegradability, appropriate characterization in terms of biosafety issues, large-scale and cost-effective synthesis, nanodrug development beyond chemotherapeutic drugs and enhanced performance efficiency. In particular, performance efficiency is the most critical issue that needs to be improved via exploring specific targeting options, sequential delivery of multiple drugs and other advanced means.

REFERENCES

1. Bobo, D., Robinson, K. J., Islam, J., Thurecht, K. J. and Corrie, S. R. 2016. Nanoparticle-based medicines: A review of FDA-approved materials and clinical trials to date. *Pharmaceutical Research*, 33, 2373–2387.

2. Caster, J. M., Patel, A. N., Zhang, T. and Wang, A. 2017. Investigational nanomedicines in 2016: A review of nanotherapeutics currently undergoing clinical trials. *Wiley Interdisciplinary Reviews. Nanomedicine and Nanobiotechnology*, 9, 1–18.
3. Sainz, V., Conniot, J., Matos, A. I., Peres, C., Zupancic, E., Moura, L., Silva, L. C., Florindo, H. F. and Gaspar, R. S. 2015. Regulatory aspects on nanomedicines. *Biochemical and Biophysical Research Communications*, 468, 504–510.
4. Wolfram, J., Zhu, M., Yang, Y., Shen, J., Gentile, E., Paolino, D., Fresta, M. et al. 2015. Safety of nanoparticles in medicine. *Current Drug Targets*, 16, 1671–1681.
5. Dick, C. A., Brown, D. M., Donaldson, K. and Stone, V. 2003. The role of free radicals in the toxic and inflammatory effects of four different ultrafine particle types. *Inhalation Toxicology*, 15, 39–52.
6. Sharma, H. S. and Sharma, A. 2007. Nanoparticles aggravate heat stress induced cognitive deficits, blood–brain barrier disruption, edema formation, and brain pathology. *Progress in Brain Research*, 162, 245–273.
7. Di Gioacchino, M., Petrarca, C., Lazzarin, F., Di Giampaolo, L., Sabbioni, E., Boscolo, P., Mariani-Costantini, R. and Bernardini, G. 2011. Immunotoxicity of nanoparticles. *International Journal of Immunopathology and Pharmacology*, 24, 65S–71S.
8. Radomska, A., Leszczyszyn, J. and Radomski, M. W. 2016. The nanopharmacology and nanotoxicology of nanomaterials: New opportunities and challenges. *Advances in Clinical and Experimental Medicine*, 25, 151–162.
9. Weissig, V., Pettinger, T. K. and Murdock, N. 2014. Nano pharmaceuticals (part 1): Products on the market. *International Journal of Nanomedicine*, 9, 4357–4373.

CHAPTER 7

Common Issues Faced in Preparation of Functional Nanoparticles and Guidelines to Solve Them

7.1 INTRODUCTION

Although many methods are reported for synthesis and functionalization of nanoparticles, only the most reproducible methods are selected and summarized in Chapters 3 through 5. However, it is still impossible to reproduce many of them for a graduate/researcher who is new in this field. In contrast, experienced researchers can reproduce most of them, particularly the methods they routinely use. This is because scientific research in this interdisciplinary area requires a great deal of expertise in different scientific disciplines, and mistake/failures can be minimized with the gain of experience. This chapter summarizes some common questions/issues/mistakes and provides guidelines to solve them.

7.2 COMMON ISSUES IN NANOPARTICLE SYNTHESIS AND PURIFICATION

7.2.1 How Does Citrate Control Au Nanoparticle Size?

Citrate can act as a capping agent for nanoparticles at room temperature. However, at high temperatures, it can act as a reducing agent as well.[1] Thus, if a citrate-borohydride mixture is added to gold salt at room temperature, borohydride acts as a reducing agent and citrate acts as a capping agent. As borohydride is a strong reducing agent, it induces more nucleation, and citrate caps the particles, which results in a 3–4 nm size. However, as citrate is added to gold salt at 100°C, citrate acts as weaker reducing agent as well as a capping agent, which results in 12–20 nm particles. Under this condition, size can be further increased up to 50 nm by lowering the citrate concentration.

7.2.2 Quantum Dot Size Control: Training Is a Must

Size-controlled synthesis of QDs is extremely important, as their emission color depends on size. For example, the emission property of CdSe is highly sensitive to size. Typically, 2-nm CdSe emits blue, 3-nm CdSe emits green, 4-nm CdSe emits yellow and 6-nm CdSe emits red. As the synthesis is done at high temperatures via the controlled nucleation-growth approach, a slight change of reaction conditions can affect the quality of CdSe. In particular, quick injection of the Se precursor and rapid cooling of the reaction mixture to stop CdSe growth are two critical steps that require a great deal of training.[2,3]

7.2.3 Why Do We Need ZnS Shelling around CdSe QDs?

ZnS shelling around CdSe nanoparticles protects the CdSe fluorescence during extensive surface chemistry.[3] ZnS shelling around CdSe requires removal of excess surfactant, optimum growth temperature and step-by-step addition of Zn and S precursors. As CdSe can grow or dissolve and ZnS nanoparticles can form separately under this condition, training and experience are essential.

7.2.4 Iron Oxide Nanoparticle Size Control: Make Sure of Reaction Conditions

The size of iron oxide nanoparticles can be varied from 5 to 40 nm, depending on the reaction temperature, ratio of fatty acids to fatty amines, chain length of fatty acids/amines and reaction time.[4] In addition, too long a reaction time can lead to poor quality of nanoparticles, so select the condition as per your size requirement.

7.2.5 Gold Nanorod Synthesis: Make Sure from Ultraviolet-Visible Spectra That Nanorods Are Formed with High Yield

There are many good methods for Au nanorod synthesis. However, the quality of the nanorods depends on the seed quality, cetyl trimethyl ammonium bromide surfactant concentration, Ag ion concentration and seed-to-gold salt ratio.[5-8] Ideally, use CTAB to prepare Au seeds via borohydride reduction and make sure that all borohydrides are consumed (wait 2–4 h after preparing seeds). Check the UV-visible spectra of the nanorod solution and see if the plasmon peak at 500 nm is significantly weaker than the plasmon peak at 700–1000 nm. This will ensure that there are very few spherical nanoparticles.

7.2.6 Colloidal Stability: Citrate versus CTAB as Capping Agent

Citrate-capped Au/Ag nanoparticles are susceptible to aggregation, while CTAB capped particles are less so. This is because citrate is a weaker capping agent than CTAB. Therefore, handle citrate-capped nanoparticles carefully, as they can precipitate in the presence of a small amount of contaminant.

7.2.7 Purification of Nanoparticles from Excess Surfactant: A Must for Successful Coating

Nanoparticles are usually synthesized in the presence of large excesses of surfactants. Such excess surfactants need to be removed before they are subjected to coating chemistry. For example, hydrophobic quantum dots/iron oxides are synthesized in the presence of large excesses of fatty amine/fatty acid surfactants and only a small fraction of them are capped on the nanoparticle surface. These surfactants should be removed via methanol/ethanol/acetone-based precipitation and a chloroform/toluene/cyclohexane-based redispersion approach. Gold nanorods are synthesized in the presence of cetyltrimethyl ammonium bromide surfactant. These surfactants need to be removed via high-speed centrifuge-based precipitation of nanorods and redispersion of nanorods in fresh water.

7.2.8 Nanoparticle Purification: Selecting Solvents Is Critical

Hydrophobic nanoparticles (such as iron oxide, QD, TiO_2, hydroxyapatite) are capped with fatty acids and fatty amines and usually synthesized using octadecene/fatty acid/fatty amine solvents. The first round of purification requires dissolving the particles in chloroform and precipitating the particles by acetone. From the second round onward, chloroform/toluene/hehaxane-based dispersion and methanol/ethanol-based precipitation are ideal. However, other solvents may be used depending on the situation. Always use the minimum methanol/ethanol to precipitate the particles, and you may use mild heating to facilitate the precipitation.

7.2.9 How Many Times Should I Purify Nanoparticles?

Most capping surfactants are weakly adsorbed (except thiols). Thus, multiple washes can lead to complete removal of surfactants, with the formation of insoluble particles. Therefore, just stop before the last step of purification that leads to insoluble particles!

7.2.10 Purification of Fatty Amine/Acid-Capped Au/Ag Nanoparticles: Do Once

Purification of fatty acid/amine-capped Au/Ag is critical, as they are susceptible to agglomeration. Add a drop of oleic acid (if not used during synthesis), as this is a better capping agent than other fatty amines/acids, and ion pairing with a fatty amine can further enhance the capping property. Do only one time precipitation-redispersion for their purification.

7.2.11 Poly Lactic-co-Glycolic Acid Nanoparticle Size Control

PLGA nanoparticles are large (>100 nm) and very polydispersed in size. Compared to inorganic nanoparticles, size control of PLGA nanoparticles is difficult. The use of a stabilizing polymer/surfactant can decrease the size a little bit. Different-sized particles may be separated by centrifuging the colloidal particles at different speeds.

7.3 COMMON ISSUES IN COATING CHEMISTRY

7.3.1 Silica-Coated Nanoparticles: Thin Shell Is Better, but Be Careful about Their Colloidal Stability in Physiological Conditions

The thin silica shell around nanoparticles is critical for many biomedical applications. Although there are many methods of silica coating, most of them produce a thick silica shell with >10 nm thickness. One advantage of a thick silica shell is that the resultant particles have good colloidal stability. However, when the overall size (or hydrodynamic size) becomes >50 nm, this larger size has limited biomedical applications. The method described in Sections 4.2.1 through 4.2.3 produces a thin silica shell (<10 nm) with a resultant hydrodynamic size of <10 nm.[9] The use of trimethoxy silane (instead of tetramethoxy-/tetraethoxysilane) and precipitation of particles from the medium immediately after silica shell formation induce such thin silica shell formation.

However, the colloidal stability of thin silica-coated nanoparticles is very sensitive to solution composition. For example, nanoparticles usually precipitate in a phosphate buffer solution or under long-term preservation.[10] Thus, silica-coated particles need to be used within a few hours of their preparation, or they require additional surface chemistry (e.g., hyperbranched polyglycerol coating; see Section 4.7) to improve their colloidal stability.[11]

7.3.2 Polyacrylate Coating: The Nature of Acrylate Monomer Is Important

At the initial process of polyacrylate coating, the original capping ligands at the nanoparticle are replaced by acryl monomers, followed by polyacrylate coating. This type of ligand exchange process is essential for effective coating around nanoparticles. However, the nature of acryl monomers dictates such ligand exchange processes. In particular, amine-containing acryl monomers are the most efficient to replace the original capping ligand, and PEG-acrylate monomers are the least efficient in such a ligand exchange. Thus, polyacrylate coating with only PEG-acrylate is usually unsuccessful. However, if more than one acrylate monomer is used and at least one monomer is able to replace the original capping ligand, polyacrylate coating would be successful.[12–15] Thus, the choice of acrylate monomers is critical for polyacrylate coating.

7.3.3 Polyacrylate Coating: An Oxygen-Free Atmosphere Is Critical

Polyacrylate coating requires an oxygen-free atmosphere, as free radicals are involved in coating processes, and the presence of traces of oxygen can inhibit polymerization as free radicals are consumed by oxygen. Therefore, purge the solution with nitrogen/argon to remove all the dissolved oxygen before the addition of persulfate and make sure that the reaction medium is free from oxygen during the whole coating process.[12–14]

7.3.4 Polyacrylate Coating: Easy for <10 nm Particles but Difficult for Larger Particles

Reverse micelles are used as a reaction medium where both the hydrophobic/hydrophilic nanoparticles and acrylate monomers can be dispersed/dissolved. However, larger nanoparticles of >10 nm are difficult to disperse in reverse micelles, so the method is most effective for nanoparticles with size <10 nm.[12-15] In particular, magnetic nanoparticles with size >25 nm are difficult to disperse in reverse micelles and result in aggregated particles after polyacrylate coating. However, anisotropic nanoparticles (e.g., TiO_2 nanorods, hydroxyapatatite nanorods, doped TiO_2 nanoparticles, Au nanorods) can be dispersed in reverse micelles due to the high surface area.[16-18]

7.3.5 Polyacrylate Coating: Make Sure Nanoparticles Are Dispersed before the Polymerization Starts

Nanoparticles may precipitate before polyacrylate coating. This can be seen as visible precipitation during mixing or a change of color (pink to blue for Au nanoparticles, yellow to blue for Ag nanoparticles) and the appearance of turbidity. This means coating will occur around aggregated nanoparticles and result in insoluble/larger particles. In such a case, restart and redesign the method. If you first dissolve the acrylate monomers in reverse micelles and then add nanoparticles, the chance of nanoparticle aggregation will be minimal. However, the final result will depend on the nanoparticle size and the nature of the acrylate monomers.

7.3.6 Lipophilic Polymer Coating: Use the Minimum Amount of Polymer

Free polymers need to be removed after coating so that the conjugation chemistry can be very effective. However, free polymers are difficult to remove by centrifuge, particularly when the particle size is <10 nm. Acetone-base precipitation and redispersion in fresh water can be exercised to remove free polymers, but this method is very inefficient. Thus, try to use the minimum amount of polymer for coating so that the percentage of free polymers will be low.[19,20]

7.4 COMMON ISSUES IN CONJUGATION CHEMISTRY

7.4.1 N-Hydroxysuccinimide/Maleimide Reagents: Poor Chemical Stability

N-hydroxysuccinimide (NHS) and maleimide-based reagents are unstable at room temperature and reactive to moisture. Usually, they are preserved inside a freezer at a low temperature and protected from moisture, so always keep them under protected conditions and take them out of the freezer when you intend to use them. If you want to use them, open the container only after attaining room temperature. After use, quickly close the container and put back in the freezer.

7.4.2 NHS/Maleimide Reaction: Use Anhydrous Solvent

NHS and maleimide react with water, so the solvents used for conjugation reactions should be free from traces of water. Preparation of anhydrous dimethylformide/DMSO requires special skills, and you may take help from a synthetic organic chemist.

7.4.3 1-Ethyl-3-(3-(Dimethylamino)propyl) Carbodiimide Coupling: Use Freshly Prepared EDC/NHS Reagents

Always use a freshly prepared 1-ethyl-3-(3-(dimethylamino)propyl) carbodiimide (EDC) solution for conjugation reactions. This is because EDC reacts with water. Always use EDC solution within minutes, and never use an old batch of EDC solution.[21] The same is also true for NHS reagents.

7.4.4 Cyanoborohydride: Do Not Use for Au Nanoparticle-Based Conjugation

Cyanoborohydride is commonly used to reduce imine bonds formed during the Schiff-base reaction between aldehyde and primary amine. In addition, it is also used in carbohydrate conjugation with primary amine-terminated nanoparticles. However, cyanide is known to react with gold in the presence of oxygen. Thus, gold nanoparticles will be dissolved in the presence of cyanoborohydride.[22] Therefore, if you are using gold nanoparticles, try to avoid cyanoborhydride (use borohydride, if it works).

7.5 COMMON ISSUES FOR PURIFICATION OF CONJUGATED NANOPARTICLES

7.5.1 Centrifuge: Most Convenient but Ineffective for Smaller Nanoparticles

After the conjugation reaction occurs, the bioconjugated nanoparticles need to be separated from unreacted reagents and biomolecules. High-speed centrifuge (10,000–25,000 rpm) is the most useful and convenient approach to separate bioconjugated nanoparticles from unreacted/unbound/free biomolecules. Usually, water-soluble biomolecules stay as a supernatant and nanoparticles precipitate from the solution. Next, precipitated nanobioconjugates are isolated and redispersed in fresh water. The method may be repeated two to three times for complete separation of unbound biomolecules. However, this approach cannot be used for <10 nm nanoparticles, as they do not precipitate, even at high-speed centrifuge.

7.5.2 Dialysis: Most Convenient but Time Consuming and Requires Costly Membrane

Dialysis is the most commonly adopted approach for separation of unbound biomolecules. This is particularly useful when the molecular weight of biomolecules is less than 1000 Da. However, dialysis typically requires 24 h or more and becomes inefficient for separation of high-molecular-weight biomolecules such as proteins/peptides/antibodies.

7.5.3 Salt-Induced Precipitation of Polyacrylate-Coated Nanobioconjugates

Salt-induced precipitation of polyacrylate-coated nanobioconjugates can be an alternative approach for purification. This approach strictly depends on the nature of functional groups and the size of nanoparticles. Typically, solid Na_2HPO_4 is gradually added and dissolved in the aqueous solution of nanobioconjugates until the complete precipitation of particles. Next, collect the precipitate and dissolve in fresh water. This salt-based precipitation and the aqueous redispersion steps can be repeated two to three times for efficient separation.

7.5.4 Acetone-Based Precipitation of Polymer-Coated Nanobioconjugates

This approach can be used for polyaspartic acid or polymaleic anhydride-coated nanobioconjugates. In this approach, acetone is added to an aqueous solution of nanobioconjugates until particle precipitation is observed or a cloudy solution appears. Next, isolate the particles via centrifuge and redisperse particles in fresh water. The precipitation-redispersion may be repeated two to three times. Compared to dialysis, this approach is rapid, but particle precipitation efficiency is often poor in many cases. In addition, traces of acetone should be removed at the last stage for any cell labeling experiments, as acetone can induce cytotoxicity.

7.5.5 Size Exclusion Chromatography/Gel Permeation Chromatography: Costly and Requires Training

This is the most powerful approach for separation of nanobioconjugate from the mixture of high-molecular-weight polymers/proteins/antibodies. However, it needs the appropriate chromatography column and expertise. In particular, this approach can be adapted for <50 nm nanobioconjugates and separation of biomolecules of 1000–100,000 Da molecular weight.

REFERENCES

1. Ji, X., Song, X., Li, J., Bai, Y., Yang, W. and Peng, X. 2007. Size control of gold nanocrystals in citrate reduction: The third role of citrate. *Journal of the American Chemical Society*, 129, 13939–13948.

2. Li, J. J., Wang, Y. A., Guo, W. Z., Keay, J. C., Mishima, T. D., Johnson, M. B. and Peng, X. G. 2003. Large-scale synthesis of nearly monodisperse CdSe/CdS core/shell nanocrystals using air-stable reagents via successive ion layer adsorption and reaction. *Journal of the American Chemical Society*, 125, 12567–12575.

3. Cao, H., Ma, J., Huang, L., Qin, H., Meng, R., Li, Y. and Peng, X. 2016. Design and synthesis of antiblinking and antibleaching quantum dots in multiple colors via wave functional confinement. *Journal of the American Chemical Society*, 138, 15727–15735.

4. Jana, N. R., Chen, Y. and Peng, X. 2004. Size- and shape-controlled magnetic (Cr, Mn, Fe, Co, Ni) oxide nanocrystals via a simple and general approach. *Chemistry of Materials*, 16, 3931–3935.

5. Jana, N. R. 2005. Gram-scale synthesis of soluble, near-monodisperse gold nanorods and other anisotropic nanoparticles. *Small*, 1, 875–882.

6. Sau, T. K. and Murphy, C. J. 2004. Seeded high yield synthesis of short Au nanorods in aqueous solution. *Langmuir*, 20, 6414–6420.

7. Nikoobakht, B. and El-Sayed, M. A. 2003. Preparation and growth mechanism of gold nanorods (NRs) using seed-mediated growth method. *Chemistry of Materials*, 15, 1957–1962.

8. Chang, H.-H. and Murphy, C. J. 2018. Mini gold nanorods with tunable plasmonic peaks beyond 1000 nm. *Chemistry of Materials*, 30, 1427–1435.

9. Jana, N. R., Earhart, C. and Ying, J. Y. 2007. Synthesis of water-soluble and functionalized nanoparticles by silica coating. *Chemistry of Materials*, 19, 5074–5082.

10. Basiruddin, S. K., Saha, A., Pradhan, N. and Jana, N. R. 2010. Advances in coating chemistry in deriving soluble functional nanoparticle. *The Journal of Physical Chemistry C*, 114, 11009–11017.

11. Panja, P., Das, P., Mandal, K. and Jana, N. R. 2017. Hyperbranched polyglycerol grafting on the surface of silica-coated nanoparticles for high colloidal stability and low nonspecific interaction. *ACS Sustainable Chemistry & Engineering*, 5, 4879–4889.

12. Saha, A., Basiruddin, S. K., Maity, A. R. and Jana, N. R. 2013. Synthesis of nanobioconjugates with a controlled average number of biomolecules between 1 and 100 per nanoparticle and observation of multivalency dependent interaction with proteins and cells. *Langmuir*, 29, 13917–13924.

13. Chakraborty, A. and Jana, N. R. 2015. Clathrin to lipid raft-endocytosis via controlled surface chemistry and efficient

perinuclear targeting of nanoparticle. *The Journal of Physical Chemistry Letters*, 6, 3688–3697.

14. Dalal, C., Saha, A. and Jana, N. R. 2016. Nanoparticle multivalency directed shifting of cellular uptake mechanism. *The Journal of Physical Chemistry C*, 120, 6778–6786.

15. Chakraborty, A., Dalal, C. and Jana, N. R. Colloidal nanobioconjugate with complementary surface chemistry for cellular and subcellular targeting. *Langmuir*, 34, 13461–13471.

16. Basiruddin, S. K., Maity, A. R., Saha, A. and Jana, N. R. 2011. Gold-nanorod-based hybrid cellular probe with multifunctional properties. *The Journal of Physical Chemistry C*, 115, 19612–19620.

17. Das, P. and Jana, N. R. 2016. Length-controlled synthesis of calcium phosphate nanorod and nanowire and application in intracellular protein delivery. *ACS Applied Materials & Interfaces*, 8, 8710–8720.

18. Biswas, A., Chakraborty, A. and Jana, N. R. 2018. Nitrogen and fluorine codoped, colloidal TiO$_2$ nanoparticle: Tunable doping, large red-shifted band edge, visible light induced photocatalysis, and cell death. *ACS Applied Materials & Interfaces*, 10, 1976–1986.

19. Ali, H., Bhunia, S. K., Dalal, C. and Jana, N. R. 2016. Red fluorescent carbon nanoparticle-based cell imaging probe. *ACS Applied Materials & Interfaces*, 8, 9305–9313.

20. Debnath, K., Mandal, K. and Jana, N. R. 2016. Phase transfer and surface functionalization of hydrophobic nanoparticle using amphiphilic poly(amino acid). *Langmuir*, 32, 2798–2807.

21. Pradhan, N., Jana, N. R. and Jana, N. R. 2018. Inhibition of protein aggregation by iron oxide nanoparticles conjugated with glutamine- and proline-based osmolytes. *ACS Applied Nano Materials*, 1, 1094–1103.

22. Basiruddin, S. K., Maity, A. R. and Jana, N. R. 2012. Glucose/galactose/dextran-functionalized quantum dots, iron oxide and doped semiconductor nanoparticles with <100 nm hydrodynamic diameter. *RSC Advances*, 2, 11915–11921.

Index

Printed and bound by CPI Group (UK) Ltd, Croydon, CR0 4YY

21/10/2024

01777086-0003